Basics of
NANO BIOTECHNOLOGY

Basics of
NANO BIOTECHNOLOGY

Sidharth Baliyan

ANMOL PUBLICATIONS PVT. LTD.
NEW DELHI-110 002 (INDIA)

ANMOL PUBLICATIONS PVT. LTD.

Regd. Office: 4360/4, Ansari Road, Daryaganj,
New Delhi-110002 (India)
Tel.: 23278000, 23261597, 23286875, 23255577
Fax: 91-11-23280289
Email: anmolpub@gmail.com
Visit us at: www.anmolpublications.com

Branch Office: No. 1015, Ist Main Road, BSK IIIrd Stage
IIIrd Phase, IIIrd Block, Bengaluru-560 085 (India)
Tel.: 080-41723429 • Fax: 080-26723604
Email: anmolpublicationsbangalore@gmail.com

Basics of Nano Biotechnology

First Edition, 2011

ISBN 978-81-261-4903-2

PRINTED IN INDIA

Printed at Suman Printers, Delhi

Contents

Contents

Preface

Nanobiotechnology is the branch of nanotechnology with biological and biochemical applications or uses. Nanobiotechnology involves the studies of existing elements of nature in order to fabricate new devices. The term bionanotechnology is often used interchangeably with nanobiotechnology, though a distinction is sometimes drawn between the two. If the two are distinguished, nano biotechnology usually refers to the use of nanotechnology to further the goals of biotechnology, while bionanotechnology might refer to any overlap between biology and nanotechnology, including the use of biomolecules as part of or as an inspiration for nanotechnological devices.

Application of nanotechnology in biomedical sciences will provide information with unprecedented precision and sensitivity, which will not only provide much deeper understanding of biosystems but also lead to the development of new revolutionary modalities of biomolecular manufacturing, early diagnostics, medical treatment, and disease prevention beyond the cellular level to that of DNA and individual proteins, the building blocks of the life process. The long-term goal of Nanobiotechnology is to contribute to the information transfer and explosive growth of this important research area. Nanobiotechnology is a rapidly advancing area of scientific and technological opportunity that applies the tools and processes of nano/microfabrication to build devices for studying biosystems. Researchers learn from biology to create new micro-nanoscale devices to better understand life processes at the nanoscale.

Author

Chapter 1

Introduction

NANO-BIOTECHNOLOGY

Nanotechnology (prefix 'nano' means dwarf in Greek) is an amalgamation of science and engineering that has effected the measurment in the lowest units (nanometre - billionth of a metre). Nanotechnology is a valuable contribution of the profound resear ch by scientists and engineers to create new systems with novel functionalities. This gives nanotechnology greater share of control over a wide range of usage in applications right from high-density data to detecting DNA sequence to improved taste and wider variety of fruits and vegetables. On the commercial front, nanotechnology has enhanced every field in science, engineering, and technology; a phenomenal advantage, thus giving high efficiency and profitability.

Biotechnology is the application of biological research techniques to develop products that cater to the need of living beings.Nano-biotechnology, as the name suggests, is a mariage of nanotechnology and biotechnology. Each of these technologies has a tremendous impact by itself and therefore when they complement each other it results in efficacious fourth dimensional solutions.

Nano' is the present buzzword among science, technology and even business community. But is it a mere buzzword or has some real practical applications? For the answer to this question, we will have to travel in future. But, at the present times, all leading organizations have identified nanotechnology as their next frontier of research.

Nanotechnology is the technology working at extremely tiny scales (of nanometre dimensions). When you work at this level, you can manipulate atoms, influence chemical reactions and do many other unimaginable things with extremely high precision. It has been proposed that nanotechnology may lead to the production of nanobots which will be able to travel inside the body, interact with each other and even replicate. But, the design and engineering of these nanobots will be extremely tough and demanding. Does it mean that we will have to wait forever to see these nanobots in action?

The answer to the preceding question is simple: No! Nature has everything which we may desire in our so called nanobots. Living organisms can travel inside the body, generate their own power, communicate with each other and best of all, they can replicate. But, the one problem with living organisms which prevents their direct use as nanobots is that they are not able to interact with digital computers, which is a must for nanobots as they need to transfer information between human body and a computer. If nanotechnology can engineer silicon robots and biotechnology can provide replicating organisms, the natural way to realise both functions is to combine both of these fields. Voila! We have Nanobiotechnology.

Nanobiotechnology is the merger of two distant fields of nanotechnology and biotechnology. It uses the concepts, components and processes from both of these fields to create new and novel applications. Already, basic research has began which is well demonstrated by the fact that ATP powered nano-motors have shown to be successfully working. Even though the commercial applications are still unfeasible, research into this field can yield surprises to what we believe is limits of technology.

The potential of Nanobiotechnology is immense. In future, it may be used to engineer nanobots which can act as personalized doctors acting round the clock. Or it may be used to create tiny cell-bombs which seek and destroy tumor cells while making sure that healthy cells are unaffected. Apart from medical applications, this technology may be employed to harvest energy directly from photosynthetic bacteria and convert it into electricity.

But before Nanobiotechnology realises its full potential, there are many hurdles to overcome. First and foremost is offered by our body itself. The vertebrate immune system battles most of the foreign particles called antigens (such as pathogens, etc.) by producing antibodies which render them inactive. Our body's immune system will also classify nano-machines inside the body as foreign particles (which they actually are), hence these nano-machines may not work in our bodies unless some special modifications are made.

Ethical and moral issues also haunt this nascent field. Is it ethical to put foreign particles inside the human body? The horror of self-replicating nano-machines going wild is also a debatable issue. This added to fact that since these nano-machines will have biological components, they will also be undergoing natural selection from evolution, which can result in new entities capable of replicating and harming like viruses, bacteria or in their own novel ways.

Once all these issues get fixed up, expect your life to be radically changed. Nanobiotechnology may bring immense paradigm shift that we may wonder that how did we live without it! Expect Nanobiotechnology to be for 21st century what Electronics was for 20th century and what Renaissance was for 15th century.

Nano technology is a relatively new technology which came into its own around the year 2000, Nano technology is a step into the future of miniaturization. In today's world smaller is better, examples of this are seen with electronics equipment. And the best example of this? A few years ago home computers were big and bulky, so were monitors.

Today we have slim line wafer thin monitors with flat screens and much slimmer computer bases with more power than ever before, due to the smaller circuits boards inside. The word Nano technology derives from the word Nano metre which is one thousandth of a micron, which is roughly the size of a single molecule.

Nano technology is the study, manufacture and

manipulation of any small structure or machine that can be made of as little as one molecule. Nano technology has only been made possible by the advancement in microscopic tools designed for looking at and allowing manipulation of single molecules.

For those studying molecular biology, Nano biotechnology is a major breakthrough, the molecular biologist joins forces with the Nano technologist to get a better understanding and gain access to the nanostructures and Nano machines which were designed by over 4 billion years of engineering, cell machinery and biological molecules.

Today Nano technologists are set to accomplish many different goals and achieve many things that wouldn't otherwise be possible by exploiting biological molecules and cell processes. At one time silicon scaffolding would be used as the basis for Nano structures however DNA's ladder structure provides Nano biotechnologists with the perfect natural framework from which to assemble Nano structures, with its highly specific bonding properties it is the perfect base for bringing together atoms in a predictable pattern to create a Nano structure.

Nano technologists have also come to rely on the biological molecules self-assembling properties when it comes to the creation of Nano structures, such as the lipids which are able to form liquid crystals spontaneously.

DNA is not only being put to good use in building Nano structures but is also a very essential component when it comes to building Nano machines, DNA which is an information storage molecule could very well in the future serve as the basis for the next generation of computers.

As mentioned earlier microprocessors and microcircuits have already shrunk over the last few years allowing smaller machinery, it is quite possible they will in the future shrink to Nano processors and Nano microcircuits.

This is not as far fetched as it might sound as DNA molecules can be mounted onto microchips and may eventually come to replace the microchips with electron flow channels etched in silicon. Biochips such as theses are DNA based processors that can take advantage of and use DNA's extraordinary capacity to store information.

Biochips are able to exploit the properties of DNA in order to solve complex computational problems, this put very simple means that they are used to do math. Scientific studies have show us that 1,000 DNA molecules were able to solve complex computational problems that would have taken over a century for a computer to solve, in just four months.

Other biological molecules are helping in many different ways as biologists strive towards being able to store and transmit more and more information in smaller places. A good example of this is the common CD that hold information and music, researchers are now using light absorbing molecules very similar to that found in our retinas to increase the storage capacity of a single CD by, a thousand fold.

NANOBIOTECHNOLOGY AT ITS BEST

Technology is turning around to help the humans by providing easier methods to keep up a healthy living. As the population in the modern times ages, scientists are finding it difficult to manage the right combination of drugs to make them effective. The existing means of intake of drugs orally or through injections are becoming moderately effective and can be extremely painful due to the non-availability of accurate diagnosis.

One of the newer methods to deliver drugs at the right place, time, and amounts is being addressed through nano-biotechnology. Scientists and engineers have come out with new systems where the human-body parametres are monitored using a semiconductor chip that is placed inside a human body.

The information recorded/acquired is then transmitted to another wrist-watch like equipment using telemetry, which, in turn, instructs the semiconductor chip to deliver the right amounts of drug to the right place.

For example: When a patient is being treated for diabetes, the semiconductor chip monitors the blood-sugar levels and passes on the information to the device outside of the body.

The external device, in turn, instructs the semiconductor chip that acts as the drug delivery system, to release the right amounts of insulin into the blood, thus causing less or no side effects to the patient.

Nanotechnology used in biotechnology is definitely a life-saving innovation that has come as a boon to mankind. The above semiconductor chip application is just a modicum of what it is capable in terms of its impact on human race. The real applications can turn around the life-cycle of human beings in the years to come.

The method we are currently studying is dielectrophoresis. This physical effect allows voltages on microfabricated electrodes (which we fabricate in the Integrated Nanosystems Research Facility) to manipulate living cells in a solution through electric fields, without ever coming into contact with the cells themselves.

This effect may be useful for on-chip preparation of biological and chemical specimens before further chemical, electric, and optical analysis occurs, presumably on the same chip. Although many years away, the integration of all these functions onto a single chip will ultimately allow cost effective mass production of a new generation of diagnostic and analytical tools.

In figure, the principle is illustrated schematically. If a polarizable object is placed in an electric field, there will be an induced positive charge on one side of the object and induced negative charge (of the same magnitude as the induced positive charge) on the other side of the object. The positive charge will experience a "pulling" force; the negative charge will experience a "pushing" force.

In a uniform field, as depicted in figure a, the "pulling" force will cancel the "pushing" force, and the net force will be zero.

However, in a non-uniform field, as depicted in figure b, the electric field will be stronger on one side of the object and weaker on the other side of the object. Hence, the "pulling" and "pushing" forces will not cancel, and there will be a net force on the object.

Uniform field
Net force = 0
Non-uniform field
Net force ≠ 0
A
B

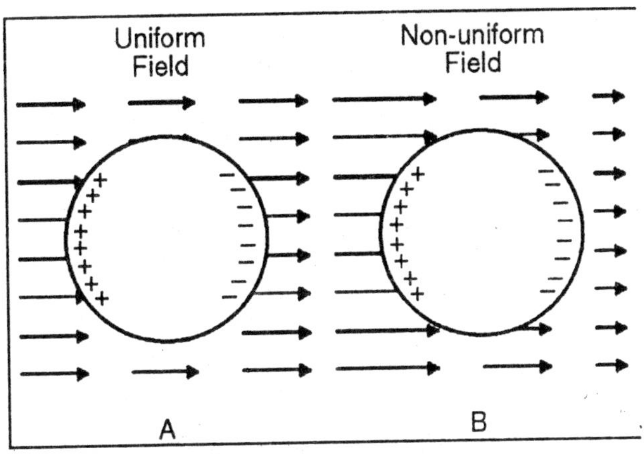

Fig. Schematic Depiction of Di-electrophoresis.

The direction of the arrows represents the direction of the electric field; the length of the arrows represents the magnitude of the electric field. In reality, the electric field lines would bend around the sphere somewhat. We neglect that effect for this simple depiction.

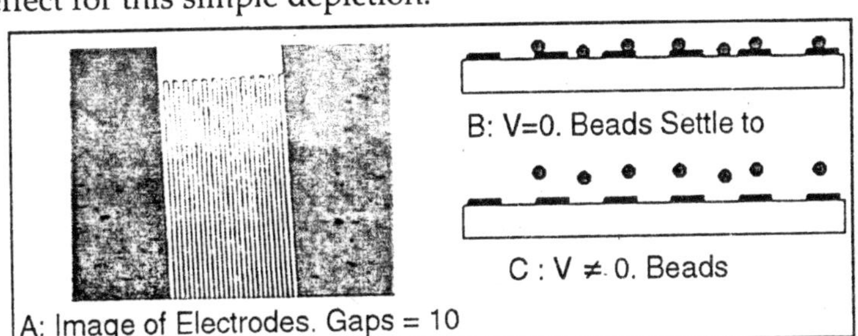

Fig. First Generation of UCI Dielectrophoresis Experiments.

A: Image of electrodes. Gaps = 10m.
B: V = 0. Beads settle to bottom.
C: V ≠ 0. Beads"levitate".

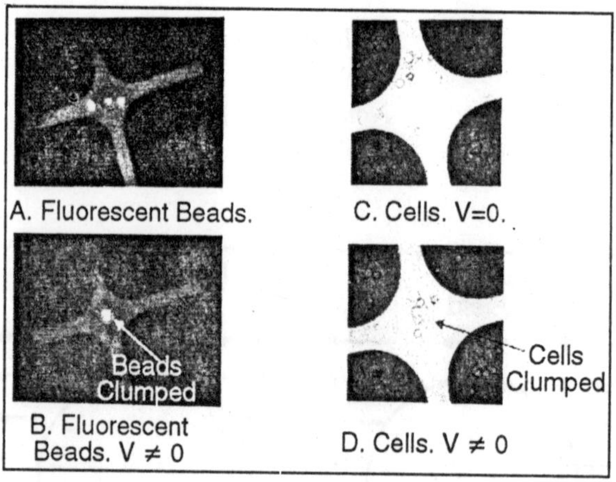

Fig. Second Generation of UCI Dielectrophoresis Experiments. There are four Electrodes which Surround the trap. The central Region is Approximately 100 m across.

A: Fluorescent beads. V= 0.

B: Fluorescent beads. V ≠ 0.

C: Cells. V=0.

D: Cells. V ≠ 0.

Beads clumped in"trap"

Cells clumped in"trap"

Nanobiotechnology is the application of nanotechnology for the study of biological and biomedical systems. One of the major areas is the utility of nanoscale systems and nanofabricated devices to guide stem cell development, mimic tissue regeneration and develop tools for cancer research.

Since stem cells are able to self-renew and differentiate into specialized cell types, so called pluripotency, they have great potential to better understand diseases, aid in drug discovery and treat diseases. The breakthrough findings that adult cells can be reprogrammed to Induced Pluripotent Stem Cells (iPSCs) provide an important cell source for regenerative medicine and drug discovery. Using a set of reprogramming factors linked to pluripotency, iPSCs can be produced from normal human cells or patient's cells with specific diseases. These iPSCs hold great promise for the study and treatment of human disease. Nanofabrication-based gene delivery of reprogramming factors into cells will facilitate generating iPSCs in a safe, efficient and effective manner.

The self-renewal and differentiation of stem cells are controlled by stem cell microenvironment which constitutes of extracellular matrices, soluble factors and neighbouring cells. The extracellular matrix is an interconnected fibrous network with nanoscale architecture. Most of the soluble factors are growth factors which are naturally occurring proteins. Cell-cell junctions play a critical role in stem cell proliferation and differentiation. It is essential to understand how stem cells interact with their microenvironment before realizing the full potential of iPSCs and other stem cells. The nanoscale interactions between stem cells and their microenvironment can be revealed by metrology tools such as scanning electron microscopy, transmission electron microscopy, atomic force microscopy and fluorescence microscopy combined with cell biology techniques. The advance in nanotechnology will provide better metrology tools as well as molecular probes to visualize cell-microenvironment interactions.

Cells and tissues in the body are three-dimensional (3-D) architectures with nanoscale interactions. Stem cells are usually studied in vitro using two-dimensional (2-D) monolayer culture. In many cases, cells grown in 2-D culture fail to recapitulate some of the key stem cell properties. Therefore, it is necessary to establish 3-D tissue models for stem cell studies. Using the tissue engineering principle, 3-D cultivation systems can be established to mimic stem cell microenvironments, propagate stem cells and differentiate them into cell types of interest. The stem cell microenvironment can be manipulated with nanofibrous matrices or surfaces with nanotopography to mimic the extracellular matrix and basement membrane, nanoparticles or micro-devices to control the release of soluble factors, cell deposition techniques to place neighbouring cells close to stem cells. The ability to reconstruct stem cell microenvironments will facilitate the understanding of stem cell-microenvironment interactions and the maintenance of stem cell properties in vitro. Micro-/Nanofabrication technologies make it possible to construct micro-tissue models for understanding stem cell differentiation and tissue morphogenesis. The drugs that regulate stem cell behaviours can be good therapeutics. In this way, the 3-D micro-tissue model provides a platform for screening therapeutics and testing drug efficacy.

A rare subtype of cancer cells shares similarities with normal stem cells which possess unlimited self-renewal potential and differentiation capacity. These cells are referred to cancer stem cells. Comparing behaviours of embryonic stem cells, iPSCs generated from patients with cancer, and cancer stem cells in 3-D cultivation system will facilitate the discovery of therapeutic target. The surrounding microenvironment of cancer cells plays a critical role in tumorigenesis and metastasis. In addition to extracellular matrix (ECM) and soluble factors, cancer cell microenvironment is a complex system composed of many cell types. Stem cells can be used to mimic the cellular components of cancer cell microenvironment. Micro-/nanotechnology can be used to construct 3-D microenvironments and deliver diagnostic and therapeutic agents. The established 3-D tumor model will bridge the conventional 2-D culture and animal experiments for cancer research, anti-cancer drug screening, and novel diagnostic and therapeutics testing.

In summary, the marriage of nanotechnology and stem cell biology will provide a new avenue to address challenges in cancer research. This exciting frontier is expected to lead to transformative change in the way we diagnose and treat cancer.

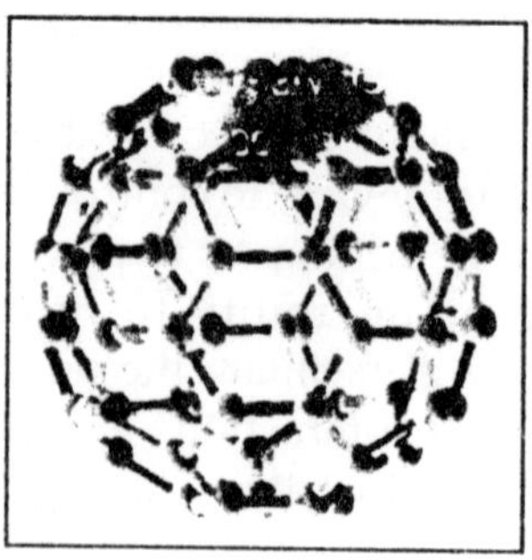

Nanobiotechnology is a new science involved in the design of tools and processes for use in the study of biosystems. These tools and processes are derived from another fairly new field called nanotechnology. Nanobiotechnology has the potential to create a new generation of medical research and practice.

NANOTECHNOLOGY

Nanotechnology is based on the use of subatomic particles as raw materials for manufacturing purposes. Subatomic particle sizes are in the nano-scale range, where one nanometre equals one-billionth of a metre. Materials reduced to nano-scale are in their purest chemical forms. In effect, nanotechnology is working with the most fundamental elements in the physical world. Manufacturing processes that begin on this level can be customised to order, meaning the materials used within the process can be created for a desired purpose.

NANOBOTS

Any manufacturing process that involves subatomic particles requires specialty tools. Nano-scaled tools are needed, as is a high degree of automation. Nanobots are the newest set of manufacturing tools designed to manipulate subatomic particles within the manufacturing process. Because nanotechnology is still in its infancy, the manufacturing of tools small enough to work with these materials is another branch in the field. Nanobots are expected to play a key role in the study of biological system processes as nanobiotechnology advances.

BIOLOGICAL SYSTEMS

Nanobiotechnology is concerned with how biological systems are able to rebuild and regenerate their own materials. Cellular functions, DNA regeneration and protein and enzyme production are some of the biosystems being examined.

Nanobiotechnology's purpose is to design tools capable of studying these systems. By understanding how these systems work, researchers hope to incorporate these designs into the workings of a nanobot that then can carry out self-directed processes.

APPLICATIONS

Areas of nanobiotechnology application in the medical field include synthetic bone materials, antimicrobial wound dressings and nanobot capsules. For instance, nanobot materials contained inside wound dressings would administer medicine in prescribed intervals and eliminate fluids discharged from the wound. Nanobot capsules are being considered as a new form of cancer treatment. Cancer-seeking nanobots would be injected into a person's bloodstream and programmed to destroy cancer cells from the inside out.

CONSIDERATIONS

Nanobiotechnology's work with subatomic materials is an uncharted course within the field of science. Subatomic particles behave differently than macro-sized materials and are more powerful as pure elements. It is unknown whether the miniscule dimension of these materials will leave residual effects in the atmosphere or if they'll create unforeseen side effects down the road. At the time of this writing, no specified regulating agencies were in charge of safety, nor were any manufacturing regulations in place to monitor the advances in this field.

FUNDAMENTAL BIOTECHNOLOGY TECHNIQUES AND APPLICATIONS

WHAT IS BIOTECHNOLOGY

Contrary to its name, biotechnology is not a single technology. Rather it is a group of technologies that share two (common) characteristics -- working with living cells and their molecules and having a wide range of practice uses that can improve our lives.

Biotechnology can be broadly defined as"using organisms or their products for commercial purposes."As such, (traditional) biotechnology has been practices since he beginning of records history. (It has been used to:) bake bread, brew alcoholic beverages, and breed food crops or domestic animals. But recent developments in molecular biology have given biotechnology new meaning, new prominence, and new potential. It is (modern) biotechnology that has captured the attention of the public. Modern biotechnology can have a dramatic effect on the world economy and society.

One example of modern biotechnology is genetic engineering. Genetic engineering is the process of transferring individual genes between organisms or modifying the genes in an organism to remove or add a desired trait or characteristic. Examples of genetic engineering are described later in this document. Through genetic engineering, genetically modified crops or organisms are formed. These GM crops or GMOs are used to produce biotech-derived foods. It is this specific type of modern biotechnology, genetic engineering, that seems to generate the most attention and concern by consumers and consumer groups. What is interesting is that modern biotechnology is far more precise than traditional forms of biotechnology and so is viewed by some as being far safer.)

HOW DOES MODERN BIOTECHNOLOGY WORK

All organisms are made up of cells that are programmed by the same basic genetic material, called DNA (deoxyribonucleic acid). Each unit of DNA is made up of a combination of the following nucleotides -- adenine (A), guanine (G), thymine (T), and cytosine (D) -- as well as a sugar and a phosphate. These nucleotides pair up into strands that twist together into a spiral structure call a"double helix."This double helix is DNA. Segments of the DNA tell individual cells how to produce specific proteins. These segments are genes. It is the presence or absence of the specific protein that gives an organism a trait or characteristic. More than 10,000 different genes are found in most plant and animal species. This total set of genes for an organism is organized into chromosomes within the cell nucleus. The process by which a multicellular organism develops from a single cell through an embryo stage into an adult is ultimately controlled by the genetic information of the cell, as well as interaction of genes and gene products with environmental factors.

When cells reproduce, the DNA strands of the double helix separate. Because nucleotide A always pairs with T and G always pairs with C, each DNA strand serves as a precise blueprint for a specific protein. Except for mutations or mistakes in the replication process, a single cell is equipped with the information to replicate into millions of identical cells. Because all organisms are made up of the same type of genetic material (nucleotides A, T, G, and C), biotechnologists use enzymes to cut and remove DNA segments from one organism and recombine it with DNA in another organism. This is called recombinant DNA (rDNA) technology, and it is one of the basic tools of modern biotechnology. rDNA technology is the laboratory manipulation of DNA in which DNA, or fragments of DNA from different sources, are cut and recombined using enzymes. This recombinant DNA is then inserted into a living organism. rDNA technology is usually used synonymously with genetic engineering. rDNA technology allows researchers to move genetic information between unrelated organisms to produce desired products or characteristics or to eliminate undesirable characteristics.

Genetic engineering is the technique of removing, modifying or adding genes to a DNA molecule in order to change the information it contains. By changing this information, genetic engineering changes the type or amount of proteins an organism is capable of producing. Genetic engineering is used in the production of drugs, human gene therapy, and the development of improved plants. For example, an"insect protection"gene (Bt) has been inserted into several crops - corn, cotton, and potatoes - to give farmers new tools for integrated pest management. Bt corn is resistant to European corn borer. This inherent resistance thus reduces a farmers pesticide use for controlling European corn borer, and in turn requires less chemicals and potentially provides higher yielding Agricultural Biotechnology.

Although major genetic improvements have been made in crops, progress in conventional breeding programmes has been slow. In fact, most crops grown in the US produce less than their full genetic potential. These shortfalls in yield are due to the inability of crops to tolerate or adapt to environmental stresses, pests, and diseases. For example, some of the world's highest yields of potatoes are in Idaho under irrigation, but in 1993 both quality and yield were severely reduced because of cold, wet weather and widespread frost damage during June. Some of the world's best bread wheats and malting barleys are produced in the north-central states, but in 1993 the disease Fusarium caused an estimated $1 billion in damage.

Scientists have the ability to insert genes that give biological defence against diseases and insects, thus reducing the need for chemical pesticides, and they will soon be able to convey genetic traits that enable crops to better withstand harsh conditions, such as drought. The International Laboratory for Tropical Agricultural Biotechnology (ILTAB) is developing transformation techniques and applications for control of diseases caused by plant viruses in tropical plants such as rice, cassava and tomato. In 1995, ILTAB reported the first transfer through biotechnology of a resistance gene from a wild species of rice to a susceptible cultivated rice variety. The transferred gene expressed resistance to Xanthomonas oryzae, a bacterium which can destroy the crop through disease. The resistant gene was transferred into susceptible rice varieties that are cultivated on more than 24 million hectares around the world.

Benefits can also be seen in the environment, where insect-protected biotech crops reduce the need for chemical pesticide use. Insect-protected crops allow for less potential exposure of farmers and groundwater to chemical residues, while providing farmers with season-long control. Also by reducing the need for pest control, impacts and resources spent on the land are less, thereby preserving the topsoil.

Major advances also have been made through conventional breeding and selection of livestock, but significant gains can still be made by using biotechnology. Currently, farmers in the U.S spend $17 billion dollars on animal health. Diseases such as hog cholera and pests such as screwworm have been eradicated. Uses of biotechnology in animal production include development of vaccines to protect animals from disease, production of several calves from one embryo (cloning), increase of animal growth rate, and rapid disease detection.

Modern biotechnology has offered opportunities to produce more nutritious and better tasting foods, higher crop yields and plants that are naturally protected from disease and insects. Modern biotechnology allows for the transfer of only one or a few desirable genes, thereby permitting scientists to develop crops with specific beneficial traits and reduce undesirable traits. Traditional biotechnology such as cross-pollination in corn produces numerous, non-selective changes. Genetic modifications have produced fruits that can ripen on the vine for better taste, yet have longer shelf lives through delayed pectin degradation.

Tomatoes and other produce containing increased levels of certain nutrients, such as vitamin C, vitamin E, and or beta carotene, and help protect against the risk of chronic diseases, such as some cancers and heart disease. Similarly introducing genes that increase available iron levels in rice three-fold is a potential remedy for iron deficiency, a condition that effects more than two billion people and causes anemia in about half that number. Most of the today's hard cheese products are made with a biotech enzyme called chymosin. This is produced by genetically engineered bacteria which is considered more purer and plentiful than it's naturally occurring counterpart, rennet, which is derived from calf stomach tissue.

In 1992, Monsanto Company successfully inserted a gene from a bacterium into the Russet Burbank potato. This gene increases the starch content of the potato. Higher starch content reduces oil absorption during frying, thereby lowering the cost of processing french fries and chips and reducing the fat content in the finished product. This product is still awaiting final development and approval.

Modern biotechnology offers effective techniques to address food safety concerns. Biotechnical methods may be used to decrease the time necessary to detect foodborne pathogens, toxins, and chemical contaminants, as well as to increase detection sensitivity. Enzymes, antibodies, and microorganisms produced using rDNA techniques are being used to monitor food production and processing systems for quality control.

Biotechnology can compress the time frame required to translate fundamental discoveries into applications. This is done by controlling which genes are altered in an organized fashion. For example, a known gene sequence from a corn plant can be altered to improve yield, increase drought tolerance, and produce insect resistance (Bt) in one generation. Conventional breeding techniques would take several years. Conventional breeding techniques would require that a field of corn is grown and each trait is selected from individual stalks of corn. The ears of corn from selected stalks with each desired trait (e.g, drought tolerance and yield performance) would then be grown and combined (cross-pollinated). Their offspring (hybrid) would be further selected for the desired result (a high performing corn with drought tolerance). With improved technology and knowledge about agricultural organisms, processes, and ecosystems, opportunities will emerge to produce new and improved agricultural products in an environmentally sound manner.

In summary, modern biotechnology offers opportunities to improve product quality, nutritional content, and economic benefits. The genetic makeup of plants and animals can be modified by either insertion of new useful genes or removal of unwanted ones. Biotechnology is changing the way plants and animals are grown, boosting their value to growers, processors, and consumers.

INDUSTRIAL BIOTECHNOLOGY

Industrial biotechnology applies the techniques of modern molecular biology to improve the efficiency and reduce the environmental impacts of industrial processes like textile, paper and pulp, and chemical manufacturing. For example, industrial biotechnology companies develop biocatalysts, such as enzymes, to synthesize chemicals. Enzymes are proteins produced by all organisms. Using biotechnology, the desired enzyme can be manufactured in commercial quantities.

Commodity chemicals (e.g., polymer-grade acrylamide) and specialty chemicals can be produced using biotech applications. Traditional chemical synthesis involves large amounts of energy and often-undesirable products, such as HCl. Using biocatalysts, the same chemicals can be produced more economically and more environmentally friendly. An example would be the substitution of protease in detergents for other cleaning compounds. Detergent proteases, which remove protein impurities, are essential components of modern detergents. They are used to break down protein, starch, and fatty acids present on items being washed. Protease production results in a biomass that in turn yields a useful byproduct- an organic fertilizer. Biotechnology is also used in the textile industry for the finishing of fabrics and garments. Biotechnology also produces biotech-derived cotton that is warmer, stronger, has improved dye uptake and retention, enhanced absorbency, and wrinkle- and shrink-resistance.

Some agricultural crops, such as corn, can be used in place of petroleum to produce chemicals. The crop's sugar can be fermented to acid, which can be then used as an intermediate to produce other chemical feedstocks for various products. It has been projected that 30% of the world's chemical and fuel needs could be supplied by such renewable resources in the first half of the next century. It has been demonstrated, at test scale, that biopulping reduces the electrical energy required for wood pulping process by 30%.

ENVIRONMENTAL BIOTECHNOLOGY

Environmental biotechnology is the used in waste treatment and pollution prevention. Environmental biotechnology can more efficiently clean up many wastes than conventional methods and greatly reduce our dependence on methods for land-based disposal.

Every organism ingests nutrients to live and produces by-products as a result. Different organisms need different types of nutrients. Some bacteria thrive on the chemical components of waste products. Environmental engineers use bioremediation, the broadest application of environmental biotechnology, in two basic ways. They introduce nutrients to stimulate the activity of bacteria already present in the soil at a waste site, or add new bacteria to the soil. The bacteria digest the waste at the site and turn it into harmless byproducts. After the bacteria consume the waste materials, they die off or return to their normal population levels in the environment.

Bioremediation, is an area of increasing interest. Through application of biotechnical methods, enzyme bioreactors are being developed that will pretreat some industrial waste and food waste components and allow their removal through the sewage system rather than through solid waste disposal mechanisms. Waste can also be converted to biofuel to run generators. Microbes can be induced to produce enzymes needed to convert plant and vegetable materials into building blocks for biodegradable plastics.

In some cases, the byproducts of the pollution-fighting microorganisms are themselves useful. For example, methane can be derived from a form of bacteria that degrades sulfur liquor, a waste product of paper manufacturing. This methane can then be used as a fuel or in other industrial processes.

HUMAN APPLICATIONS

Biotechnical methods are now used to produce many proteins for pharmaceutical and other specialized purposes. A harmless strain of Escherichia coli bacteria, given a copy of the gene for human insulin, can make insulin. As these genetically modified (GM) bacterial cells age, they produce human insulin, which can be purified and used to treat diabetes in humans. Microorganisms can also be modified to produce digestive enzymes. In the future, these microorganisms could be colonized in the intestinal tract of persons with digestive enzyme insufficiencies. Products of modern biotechnology include artificial blood vessels from collagen tubes coated with a layer of the anticoagulant heparin. Gene therapy - altering DNA within cells in an organism to treat or cure a disease - is one of the most promising areas of biotechnology research. New genetic therapies are being developed to treat diseases such as cystic fibrosis, AIDS and cancer.

DNA fingerprinting is the process of cross matching two strands of DNA. In criminal investigations, DNA from samples of hair, bodily fluids or skin at a crime scene are compared with those obtained from the suspects. In practice, it has become one of the most powerful and widely known applications of biotechnology today. Another process, polymerase chain reaction (PCR), is also being used to more quickly and accurately identify the presence of infections such as AIDS, Lyme disease and Chlamydia.

Paternity determination is possible because a child's DNA pattern is inherited, half from the mother and half from the father. To establish paternity, DNA fingerprints of the mother, child and the alleged father are compared. The matching sequences of the mother and the child are eliminated from the child's DNA fingerprint; what remains comes from the biological father. These segments are then compared for a match with the DNA fingerprint of the alleged father.

DNA testing is also used on human fossils to determine how closely related fossil samples are from different geographic locations and geologic areas. The results shed light on the history of human evolution and the manner in which human ancestors settled different parts of the world.

Chapter 2

The Molecular World

TECHNOLOGY

Nanotechnology will be a bottom-up technology, building upward from the molecular scale. It will bring a revolution in human abilities like that brought by agriculture or power machinery. It can even be used to reverse many of the changes brought by agriculture or power machinery. But we humans are huge creations with no direct experience of the molecular world, and this can make nanotechnology hard to visualize, hence hard to understand. Scientists working with moleculas face this problem today. They can often calculate how molecules will behave, but to understand this behaviour, they need more than heaps of numbers: they need pictures, movies, and interactive simulations, and so they are producing them at an ever-increasing pace. The U.S. National Science Foundation has launched a programm in"scientific visualization", in part to harness supercomputers to the problem of picturing the molecular world.

Molecules are objects that exert forces on one another. If your hands were small enough, you could grab them, squeeze them, and bash them together. Understanding the molecular world is much like understanding any other physical world: it is a matter of understanding size, shape, strength, force, motion, and the like-a matter of understanding the differences between sand, water, and rock, or between steel and soap bubbles. Today's visualization tools give a taste of what will become possible with tomorrow's faster computers and better"virtual realities,"simulated environments that let you tour a world that"exists"only as a model inside the computer. Before discussing nanotechnology and how it relates to the technologies of today, let's try to get a more concrete understanding of the molecular world by describing a simulation embedded in a scenario. In this scenario, events and technologies described as dating from 1990 or before are historically accurate; those with later dates are either projections or mere scenario elements. The descriptive details in the simulation are written to fit designs and calculations based on standard scientific data, so the science isn't fiction.

EXPLORING THE MOLECULAR WORLD

In a scenario in the last Peragraph, we saw Joel Gregory manipulating molecules in the virtual reality of a simulated world using video goggles, tactile gloves, and a supercomputer. The early twenty-first century should be able to do even better. Imagine, then, that today you were to take a really long nap, oversleep, and wake up decades later in a nanotechnological world.

In the twenty-first century, even more than in the twentieth, it's easy to make things work without understanding them, but to a newcomer much of the technology seems like magic, which is dissatisfying. After a few days, you want to understand what nanotechnology is, on a gut level. Back in the late twentieth century, most teaching used dry words and simple pictures, but now-for a topic like this-it's easier to explore a simulated world. And so you decide to explore a simulation of the molecular world.

Looking through the brochure, you read many tedious facts about the simulation: how accurate it is in describing sizes, forces, motions, and the like; how similar it is to working tools used by both engineering students and professionals; how you can buy one for your very own home, and so forth. It explains how you can tour the human body, see state-of-the-art nanotechnology in action, climb a bacterium, etc. For starters, you decide to take an introductory tour: simulations of real twentieth-century objects alongside quaint twentieth-century concepts of nanotechnology. After paying a small fee and memorizing a few key phrases (any variation of"Get me out of here!"will do the most important job), you pull on a powersuit, pocket a Talking Tourguide, step into the simulation chamber, and strap the video goggles over your eyes. Looking through the goggles, you seem to be in a room with a table you know isn't really there and walls that seem too far away to fit in the simulation chamber. But trickery with a treadmill floor makes the walk to the walls seem far enough, and when you walk back and thump the table, it feels solid because the powersuit stops your hand sharply at just the right place. You can even feel the texture of the carvings on the table leg, because the suit's gloves press against your fingertips in the right patterns as you move. The simulation isn't perfect, but it's easy to ignore the defects. On the table is (or seems to be) an old 1990s silicon computer chip. When you pick it up, as the beginners' instructions suggest, it looks like Figure.

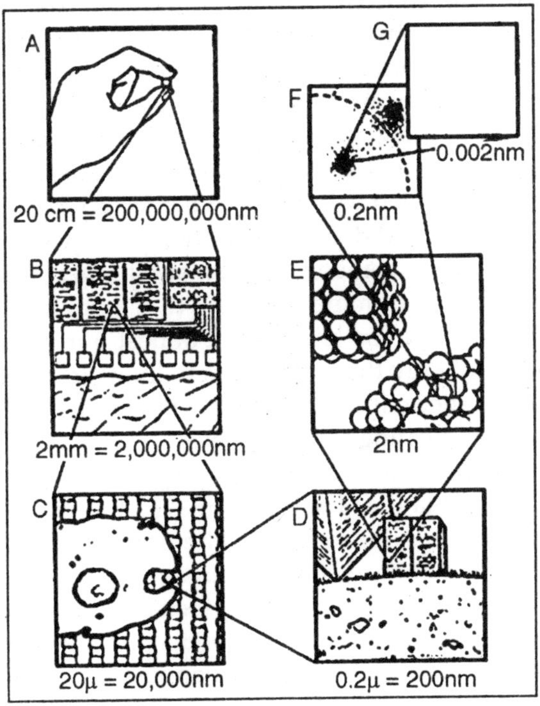

Fig. Power of Ten

Frame (A) shows a hand holding a computer chip. This is shown magnified 100 times in (B). Another factor of 100 magnification (C) shows a living cell placed on the chip to show scale. Yet another factor of 100 magnification (D) shows two nanocomputers beside the cell. The smaller has roughly the same power as the chip seen in the first view; the larger (with only the corner visible) is as powerful as mid-1980s mainframe computer. Another factor of 100 magnification (E) shows an irregular protein from the cell on the lower right, and a cylindrical gear made by molecular at top left. Taking a smaller factor of 10 jump, (F) shows two atoms in the protein, with electron clouds represented by stippling. A final factor of 100 magnification (G) reveals the nucleus of the atom as a tiny speck.

VISION AND MOTION

You feel as though you're falling towards the chip's surface, shrinking rapidly. In a moment, it looks roughly, with your thumb still there holding it. The world grows blurrier, then everything seems to go wrong as you approach the molecular level. First, your vision blurs to uselessness-there is light, but it becomes a featureless fog. Your skin is tickled by small impacts, then battered by what feel like hard-thrown marbles. Your arms and legs feel as though they are caught in turbulence, pulling to and fro, harder and harder. The ground hits your feet, you stumble and stick to the ground like a fly on flypaper, battered so hard that it almost hurts. You asked for realism, and only the built-in safety limits in the suit keep the simulated thermal motions of air molecules and of your own arms from beating you senseless.

"Stop!"gives you a rest from the suit's yanking and thumping, and"Standard settings!"makes the world around you become more reasonable. The simulation changes, introducing the standard cheats. Your simulated eyes are now smaller than a light wave, making focus impossible, but the goggles snap your vision into sharpness and show the atoms around you as small spheres. (Real nanomachinesare as blind as you were a moment ago, and can't cheat.) You are on the surface of the 1990s computer chip, between a cell and two blocky nanocomputers. Your simulated body is 50 nanometres tall, about 1/40,000,000 your real size, and the smaller nanocomputer is twice your height.

The simulation keeps bombarding you with air molecules, but the standard settings leave out the sensation of being pelted with marbles. A moment ago you were stuck tight to the ground by molecular stickiness, but the standard settings give your muscles the effective strength of steel-at least in simulation-by making everything around you much softer and weaker.

The tourguide says that the only unreal features of the simulation have to do with you-not just your ability to see and to ignore thermal shaking and bombardment, but also your sheer existence at a size too small for anything so complex as a human being. It also explains why you can see things move, something about slowing down everything around you by a factor of 10 for every factor of 10 enlargement, and by another factor to allow for your being made stronger and hence faster. And so, with your greater strength and some adjustments to make your arms, legs, and torso less sticky, you can stand, see, feel, and take stock of the situation.

MOLECULAR TEXTURE

The ground underfoot, like everything around you, is pebbly with atom-sized bumps the size of your fingertips. Objects look like bunches of transparent grapes or fused marbles in a variety of pretty but imaginary colours. The simulation displays a view of atoms and molecules much like those used by chemists in the 1980s, but with a sharper 3-D image and a better way to move them and to feel the forces they exert. Actually, the whole simulation setup is nothing but an improved version of systems built in the late 1980s-the computer is faster, but it is calculating the same things. The video goggles are better and the whole-body powersuit is a major change, but even in the 1980s there were 3-D displays for molecules and crude devices that gave a sense of touching them.

The gloves on this suit give the sensation of touching whatever the computer simulates. When you run a fingertip over the side of the smaller nanocomputer, it feels odd, hard to describe. It is as if the surface were magnetic-it pulls on your fingertip if you move close enough. But the result isn't a sharp click of contact, because the surface isn't hard like a magnet, but strangely soft. Touching the surface is like touching a film of fog that grades smoothly into foam rubber, then hard rubber, then steel, all within the thickness of a sheet of corrugated cardboard. Moving sideways, your fingertip feels no texture, no friction, just smooth bumps more slippery than oil, and a tendency to get pulled into hollows. Pulling free of the surface takes a firm tug. The simulation makes your atom-sized fingertips feel the same forces that an atom would. It is strange how slippery the surface is-and it can't have been lubricated, since even a single oil molecule would be a lump the size of your thumb. This slipperiness makes it obvious how noon-scale bearings can work, how the parts of molecular machines can slide smoothly.

But on top of this, there is a tingling feeling in your fingers, like the sensation of touching a working loudspeaker. When you put your ear against the wall of the nanocomputer, you flinch back: for a moment, you heard a sound like the hiss of a twentieth-century television tuned to a channel with no broadcast, with nothing but snow and static-but loud, painfully loud. All the atoms in the surface are vibrating at high frequencies, too fast to see. This is thermal vibration, and it's obvious why it's also called thermal noise.

GAS AND LIQUID

Individual molecules still move too quickly to see. So, to add one more cheat to the simulation, you issue the command"Whoa!", and everything around seems to slow down by a factor of ten.

On the surface, you now can see thermal vibrations that had been too quick to follow. All around, air molecules become easier to watch. They whiz about as thick as raindrops in a storm, but they are the size of marbles and bounce in all directions. They're also sticky in a magnetlike way, and some are skidding around on the wall of the nanocomputer. When you grab one, it slips away. Most are like two fused spheres, but you spot one that is perfectly round-it is an argon atom, and these are fairly rare. With a firm grip on all sides to keep it from shooting away like a watermelon seed, you pinch it between your steel-strong fingers. It compresses by about 10 per cent before the resistance is more than you can overcome. It springs back perfectly and instantly when you relax, then bounces free of your grip. Atoms have an unfamiliar perfection about them, resilient and unchanging, and they surround you in thick swarms.

At the base of the wall is a churning blob that can only be a droplet of water. Scooping up a handful for a closer look yields a swarm of molecules, hundreds, all tumbling and bumbling over one another, but clinging in a coherent mass. As you watch, though, one breaks free of the liquid and flies off into the freer chaos of the surrounding air: the water is evaporating. Some slide up your arm and lodge in the armpit, but eventually skitter away. Getting rid of all the water molecules takes too much scraping, so you command"Clean me!"to dry off.

TOO SMALL AND TOO LARGE

Beside you, the smaller nanocomputer is a block twice your height, but it's easy to climb up onto it as the tourguide suggests. Gravity is less important on a small scale: even a fly can defy gravity to walk on a ceiling, and an ant can lift what would be a truck to us. At a simulated size of fifty nanometres, gravity counts for nothing. Materials keep their strength, and are just as hard to bend or break, but the weight of an object becomes negligible. Even without the strength-enhancement that lets you overcome molecular stickiness, you could lift an object with 40 million times your mass-like a person of normal size lifting a box containing a half-dozen fully loaded oil tankers. To simulate this weak gravity, the powersuit cradles your body's weight, making you feel as if you were floating. This is almost like a vacation in an orbital theme park, walking with stickyboots on walls, ceilings, and whatnot, but with no need for antinausea medication.

On top of the nanocomputer is a stray protein molecule. This looks like a cluster of grapes and is about the same size. It even feels a bit like a bunch of grapes, soft and loose. The parts don't fly free like a gas or tumble and wander like a liquid, but they do quiver like gelatin and sometimes flop or twist. It is solid enough, but the folded structure is not as strong as your steel fingers. In the 1990s, people began to build molecular machinery out of proteins, copying biology. It worked, but it's easy to see why they moved on to better materials.

From a simulated pocket, you pull out a simulated magnifying glass and look at the simulated protein. This shows a pair of bonded atoms on the surface at 10 times magnification. The atoms are almost transparent, but even a close look doesn't reveal a nucleus inside, because it's too small to see. It would take 1,000 times magnification to be able to see it, even with the head start of being able to see atoms with your naked eye.

How could people ever confuse big, plump atoms with tiny specks like nuclei? Remembering how your steel-strong fingers couldn't press more than a fraction of the way towards the nucleus of an argon atom from the air, it's clear why nuclear fusion is so difficult. In fact, the tourguide said that it would take a real-world projectile over a hundred times faster than a high-powered rifle bullet to penetrate into the atomic core and let two nuclei fuse. Try as you might, there just isn't anything you could find in the molecular world that could reach into the middle of an atom to meddle with its nucleus. You can't touch it and you can't see it, so you stop squinting though the magnifying glass. Nuclei just aren't of much interest in nanotechnology.

PUZZLE CHAINS

Taking the advice of the tourguide, you grab two molecular knobs on the protein and pull. It resists for a moment, but then a loop comes free, letting other loops flop around more, and the whole structure seems to melt into a writhing coil. After a bit of pulling and wrestling, the protein's structure becomes obvious: It is a long chain-longer than you are tall, if you could get it straight-and each segment of the chain has one of several kinds of knobs sticking off to the side. With the multicoloured, glassy-bead portrayal of atoms, the protein chain resembles a flamboyant necklace.

This may be decorative, but how does it all go back together? The chain flops and twists and thrashes, and you pull and push and twist, but the original tight, solid packing is lost. There are more ways to go wrong in folding up the chain than there are in solving Rubik's Cube, and now that the folded structure is gone, it isn't even clear what the result should look like. How did those twentieth-century researchers ever solve the notorious"protein folding problem"? It's a matter of record that they started building protein objects in the late 1980s.

This protein molecule won't go back together, so you try to break it. A firm grip and a powerful yank straightens a section a bit, but the chain holds together and snaps back. Though unfolding it was easy, even muscles with the strength of steel-the strength of Superman-can't break the chain itself. Chemical bonds are amazingly strong, so it's time to cheat again. When you say,"Flimsy world-one second!"while pulling, your hands easily move apart, splitting the chain in two before its strength returns to normal. You've forced a chemical change, but there must be easier ways since chemists do their work without tiny superhands. While you compare the broken ends, they thrash around and bump together. The third time this happens, the chain rejoins, as strong as before. This is like having snap-together parts, but the snaps are far stronger than welded steel. Modern assembler chemistry usually uses other approaches, but seeing this happen makes the idea of molecular assembly more understandable: Put the right pieces together in the right positions, and they snap together to make a bigger structure.

Remembering the"Whoa!"command, you decide to go back to the properly scaled speed for your size and strength.

Saying"Standard settings!,"you see the thrashing of the protein chain speed up to hard-to-follow blur.

NANOMACHINES

At your feet is a ribbed, ringed cylindrical object about the size of a soup can-not a messy, loosely folded strand like the protein (before it fell apart), but a solid piece of modern nanotechnology. In the gear, everything is held in place by bonds as strong as those that strung together the beads of the protein chain. It can't unfold, and you'd have to cheat again to break its perfect symmetry. Like those in the wall of the nanocomputer, its solidly attached atomsvibrate only slightly. There's another gear nearby, so you fit them together and make the atomic teeth mesh, with bumps on one fitting into hollows on the other. They stick together, and the soft, slick atomic surfaces let them roll smoothly.

Underfoot is the nanocomputer itself, a huge mechanism built in the same rigid style. Climbing down from it, you can see through the transparent layers of the wall to watch the inner works. An electric motor an arm-span wide spins inside, turning a crank that drives a set of oscillating rods, which in turn drive smaller rods. This doesn't look like a computer; it looks more like an engineer's fantasy from the nineteenth century. But then, it is an antique design-the tourguide said that the original proposal was a piece of exploratory engineering dating from the mid-1980s, a mechanical design that was superseded by improved electronic designs before anyone had the tools to build even a prototype. This simulation is based on a version built by a hobbyist many years later.

The mechanical nanocomputer may be crude, but it does work, and it's a lot smaller and more efficient than the electronic computers of the early 1990s. It's even somewhat faster. The rods slide back and forth in a blur of motion, blocking and unblocking each other in changing patterns, weaving patterns of logic. This nanocomputer is a stripped-down model with almost no memory, useless by itself. Looking beyond it, a machine is powerful enough to compete with most computers built in 1990. This computer is a millionth of a metre on a side, but from where you stand, it looks like a blocky building looming over ten stories tall. The tourguide says that it contains over 100 billion atoms and stores as much data as a room full of books. You can see some of the storage system inside: row upon row of racks containing spools of molecular tape somewhat like the protein chain, but with simple bumps representing the 1s and 0s of computer data.

These nanocomputers seem big and crude, but the ground you're now standing on is also a computer-a single chip from 1990, roughly as powerful as the smaller, stripped-down nanocomputer at your side. As you gaze out over the chip, you get a better sense for just how crude things were a few decades ago. At your feet, on the smallest scale, the chip is an irregular mess. Although the wall of the nanocomputer is pebbly with atomic-scale bumps, the bumps are as regular as tile. The chip's surface, though, is a jumble of lumps and mounds. This pattern spreads for dozens of paces in all directions, ending in an irregular cliff marking the edge of a single transistor. Beyond, you can see other ridges and plateaus stretching off to the horizon. These form grand, regular patterns, the circuits of the computer. The horizon-the edge of the chip-is so distant that walking there from the centre would (as the tourguide warns) take days. And these vast pieces of landscaping were considered twentieth-century miracles of miniaturization?

CELLS AND BODIES

Even back then, research in molecular biology had revealed the existence of smaller, more perfect machines such as the protein molecules in cells. A simulated human cell-put here because earlier visitors wanted to see the size comparisons-its on the chip next to the smaller nanocomputer. The tourguide points out that the simulation cheats a bit at this point, making the cell act as though it were in a watery environment instead of air. The cell dwarfs the nanocomputer sprawling across the chip surface and rearing into the sky like a small mountain. Walking the nature trail around its edge would lead across many transistor-plateaus and take about an hour. A glance is enough to show how different it is from a nanocomputer or a gear: it looks organic, it bulges and curves like a blob of liver, but its surface is shaggy with waving molecular chains.

Walking up to its edge, you can see that the membrane wrapping the cell is fluid (cell walls are for stiff things like plants), and the membrane molecules are in constant motion. On an impulse, you thrust your arm through the membrane and poke around inside. You can feel many proteins bumping and tumbling around in the cell's interior fluid, and a crisscrossing network of protein cables and beams. Somewhere inside are the molecular machines that made all these proteins, but such bits of machinery are embedded in a roiling, organic mass. When you pull your arm out, the membrane flows closed behind. The fluid, dynamic structure of the cell is largely self healing. That's what let scientists perform experimental surgery on cells with the old, crude tools of the twentieth century: They didn't need to stitch up the holes they made when they poked around inside.

Even a single human cell is huge and complex. No real thinking being could be as small as you are in the simulation: A simple computer without any memory is twice your height, and the larger nanocomputer, the size of an apartment complex, is no smarter than one of the submoronic computers of 1990. Not even a bendable finger could be as small as your simulated fingers: in the simulation, your fingers are only one atom wide, leaving no room for the slimmest possible tendon, to say nothing of nerves.

For a last look at the organic world, you gaze out past the horizon and see the image of your own, full-sized thumb holding the chip on which you stand. The bulge of your thumb rises ten times higher than Mount Everest. Above, filling the sky, is a face looming like the Earth seen from orbit, gazing down. It is your own face, with cheeks the size of continents. The eyes are motionless. Thinking of the tourguide's data, you remember: The simulation uses the standard mechanical scaling rules, so being 40 million times smaller has made you 40 million times faster. To let you pull free of surfaces, it increased your strength by more than a factor of 100, which increased your speed by more than a factor of 10. So one second in the ordinary world corresponds to over 400 million here in the simulation. It would take years to see that huge face in the sky complete a single eyeblink.

Enough. At the command"Get me out!", the molecular world vanishes, and your feeling of weight returns as the suit goes slack. You strip off the video goggles-and hugely, slowly, blink.

Chapter 3

Biotechnology Techniques and Applications

BIOTECHNOLOGY TECHNIQUES

BIOTECHNOLOGY FEATURES

Biotechnology deals with micro- organisms, plants or animal cells or their enzymes to produce products and processes useful to humans. Two core techniques are involved in modern biotechnology.

These techniques include:

- Genetic engineering is modification of chemical nature of DNA/RNA and introduced them into other host organisms to change phenotypic character of host.
- Maintain growth and manipulation of the desired microbes or cells in large quantities for manufacture of biotechnological products.

Modern biotechnology techniques are based only on the isolation and manipulation of DNA these days. These techniques made it possible to detect diseases and make certain precautionary measures to protect people from these diseases. Some of the techniques are described in this article.

DNA ISOLATION

DNA isolation techniques totally depend on the experimental organism, but somehow these techniques have following characteristics.

- A treatment for opening the cells and releasing their DNA
- Method for inactivating and removing the enzymes which degrade the DNA
- Method to separate the DNA from proteins and other molecules which contaminate the DNA

There are many cheap laboratory exercises for the isolation of DNA which are also very simple. DNA is usually taken from onion, bacteria and other organisms. Heat and detergent is used to isolate the DNA. These methods break open the cell and inactivate the enzymes which degrade DNA. DNA is separated from other contaminants by using alcohol precipitation. The method of DNA isolation is more reproducible is isolation the DNA from bacteria16. There are many companies who sell kits to for DNA isolation. DNA is isolated from human body cells to detect genetic diseases in the newborn child, to analyse the forensic evidence and to study genes which are involved in causing cancer.

DNA TRANSFORMATION

Over the past couple of decades biotechnologists have succeeded in developing DNA transfer techniques. It is the only the DNA which is required for the transformation. To get the DNA, in most cases tissues are exposed to stresses which enable the plasma membrane to rupture and in other cases cell wall of the plants is used. Through this process some of the cells will contain the foreign DNA in their genome. This foreign DNA is helpful to make useful products and has been used in pharmaceuticals such as human insulin.

RESTRICTION ENZYME DIGESTION AND ANALYSIS

Restriction enzymes are actually the proteins. They cut the sequences of bases of DNA at specific places. Biotechnologists are able to reproduce cut DNA molecules into small well defined fragments through these enzymes. In biotechnology, restriction enzymes are used to make and analyse new DNA molecules. Agarose gel electrophoresis is the technique through which biotechnologists are able to make new DNA molecules.

Agarose gel makes new DNA molecules through following methods:

- Chains of sugars make agarose.
- DNA is moved in the agarose gel through electric current and this current cuts them into smaller pieces according to their size.
- Special kind of dye makes possible the identification of small DNA molecules in the gel.

POLYMERASE CHAIN REACTION

Polymerase chain reaction or PCR is a common technique in biotechnology. This technique is used for producing multiple copies of short DNA molecules.

PCR runs through following process:

- DNA molecule that will act as a template in the reaction
- Short and single stranded DNA molecules which are called as primers. Primers are the starting points for making new DNA molecule.
- An enzyme is needed which makes DNA
- Nucleotides which are the building blocks of DNA
- Thermal cycler, a machine, which subjects the reaction mixture to varying temperatures repeatedly.

DNA SEQUENCING

DNA sequencing is the most important thing in the human genome project. Many biotechnology techniques are dependent on this method. To determine the exact order of bases in the DNA molecule is this technique's purpose. Nature of protein encoded in the DNA molecule can be determined through this technique.

MICROARRAYS

It is a very useful technique of biotechnology. In this technique, expression of many genes can be analysed simultaneously. Microarrays can be made on a glass slide or a chip and they contain hundreds and thousands of parts of genes in a small section of the slide. Expressed genes can be isolated from the cells and then are detected. This detection shows that which DNAs bind to which genes and are taken from which cells.

PRINCIPLES OF BIOTECHNOLOGY

The rDNA was introduced into E. coli and manipulate them to make a number of copies.

The step for genetic engineering includes the following techniques:

- Formation of recombinant DNA
- Use of gene cloning
- Gene transfer

Tools of Recombinant Dna Technology in Biotechnology:

The key tools required for the recombinant DNA techniques are,

- Restriction enzymes
- Polymerase enzymes
- Ligases
- Vectors
- Host organism/cell

Restriction Enzymes in Biotechnology

The restriction enzymes techniques are called as"molecular scissors", act as a foundation of recombinant DNA technology.

It occurs in bacteria to provide a type of defence called as"restriction DNA technology. Restriction enzymes techniques belong to a class of enzymes called nuclease's. These nuclease's have two types. They are Exonucleases and Endonucleases.

Polymerase Enzymes in Biotechnology

Polymerase technique is an enzyme that generates RNA. In cells, RNAP is necessary one for producing RNA chains from DNA genes as the particular form and this process called transcription.

Ligases in Biotechnology

Ligase techniques can link collectively two DNA strands that have double-strand split. Single-strand break is unchanging by a different types of DNA ligase using the harmonizing strand as a process but it still requires DNA ligase.

Vector in Biotechnology

The vectors are the DNA molecules that can carry a foreign DNA segment and replicate inside the host cell. Vectors may be plasmids, bacteria phages and cosmids.

Host Organism/cell in Biotechnology

The cell is the efficient basic unit of life. It is the minimum unit of life that is classified as a living thing, and is often called the building block of life. The examples are bacteria are unicellular. The multi cellular organisms are humans and big animals. Humans have concerning 100 trillion or 10^{14} cells; a cell volume is 10 μm; a cell bunch is 1 nanogram. The largest cells are135 μm in the granule cells in the cerebellum, the smallest can be some 4 μm.

TECHNIQUES IN PLANT BIOTECHNOLOGY

Biotechnology offers great opportunities for plant breeding. New techniques for rapidly selecting or inducing the desired characteristics are being developed. The Dutch plant breeding sector remains aloof from genetic modification in plants. The aversion of the consumer, complicated legislation and the high costs of introducing GM crops and their products do not make genetic modification an attractive alternative to conventional breeding methods. Nonetheless, with the advance of technology, the distinction between genetic modification and other plant biotechnological techniques gradually blurs. In addition, such technological developments also outgrow the GMO legislation. At times it is not clear whether the products of some techniques are subject to the prevailing GMO legislation.

Consequently, an impasse has arisen between the Dutch government and breeding companies as the developer of new techniques in plant biotechnology. Companies are only prepared to further develop some innovations when it is clear whether they are subject to the GMO legislation or not. Being bound by EU legislation, the government says it can only make this judgment when an actual application is submitted. Thus a situation is created in which both parties are waiting for each other.

Advice: This advisory report, which to some degree has a informative character, discusses six new techniques: breeding agro inoculation, grafting on genetically modified rootstock, gene silencing by DNA methylation, the use of oligonucleotides, and specific mutagenesis with homologous recombination. These techniques were chosen as they are either in the early stages of commercial application or give insight into the problem at stake. For some of the discussed techniques, the important questions are whether they can be considered genetic modification and whether their products must be characterized as GMOs. In this respect, a progressive scale can be distinguished.

The products of some techniques, such as the offspring in case of reverse breeding, do not contain any novel characteristics, added sequences, mutations or other changes. In epigenetic mutants, no sequence changes are made in the genome, though there are heritable effects. In products of grafting, transgenic sequences may be absent but transgenic proteins or other transgenic molecules or induced effects can be present.

Other products, for instance those generated by the application of mutagentia coupled with oligonucleotides, do contain mutations in the genome but that production method is similar to that of organisms exempted from the legislation. Finally, some organisms are genetically modified but by a modification technique that dismisses many of the current technical-scientific objections.

The European legislation is based on the principle that when recombinant DNA techniques are used in the production of an organism, this organism is considered a GMO with changed genetic characteristics. Therefore this organism is subject to the GMO legislation. The underlying idea here is that the process of genetic modification is inherently unsafe and associated with risks.

However, with the advance of science and biotechnology, is has become possible to use recombinant DNA techniques or genetic modification in a production process, in such a way that the resulting plant or organism does not contain any added sequences or expresses other changes. An example of this are plants that are produced with the help of reverse breeding. Based on technical scientific grounds COGEM is of the opinion that such plants should not be seen as GMOs.

If current legislation implies this is not possible, COGEM recommends that they be exempt from GMO legislation. COGEM considers further the offspring of agroinoculated plants in principle not as GMOs. However, at this moment it cannot entirely be excluded that this offspring possesses unintended transgenic sequences after agro-inoculation. COGEM will conduct further research into this. Expectations are that the results of this research will be made available at the start of 2007.

As yet it is too early for a judgment on epigenetic applications and possibly related environmental risks. The stability of epigenetic changes and the underlying mechanism of heredity are unclear at this moment.

Applications are not immediately expected. Furthermore, it is uncertain whether epigenetic mutants fall within the legal scope of GMO legislation. Whether non-modified upper stem grafted on GM rootstock and their products must be subject to the GMO legislation is principally a legal and political question. However, COGEM observes that it cannot be said that there are by definition no risks to people and the environment from the upper stem (products) grafted on GM rootstock and COGEM recommends a case by case approach.

COGEM considers specific mutagenesis with oligonucleotides a form of mutagenesis. It should therefore be exempt from GMO legislation and regulations.

Targeted integration of transgenes in plants via homologous recombination falls under the denominator of genetic modification. Plants that are produced with this technique must be considered as transgene. This implies, under the current legislation, that an environmental risk analysis will always have to be performed when a transgene is thus inserted.

This report explores contemporary techniques and applications of plant biotechnology, illustrating the tremendous potential this technology has to change our world by improving the food supply. As an introductory text, its focus is on basic science and processes. It guides students from plant biology and genetics to breeding to principles and applications of plant biotechnology. Next, the text examines the critical issues of patents and intellectual property and then tackles the many controversies and consumer concerns over transgenic plants.

Inspirational autobiographical essays, written by pioneers and eminent scientists in the field today, are interspersed throughout the text. Authors explain how they became involved in the field and offer a personal perspective on their contributions and the future of the field.

This text is recommended for junior and senior-level courses in plant biotechnology or plant genetics and for courses devoted to special topics at both the undergraduate and graduate levels. It is also an ideal reference for practitioners.

Provides a thorough overview of botany and genetics for those students with minimal background in those areas, with coverage extending from very basic introductory materials to advanced techniques and applications.

BIOTECHNOLOGY IN THE MODERN WORLD

The scientific knowledge which is applied to practical ways in industry for the benefit of human beings is called as technology.

The biological sciences have recently passed through more advanced technologies in different spheres of life and activities. One of those advanced technologies is the Bio- technology.The applications of advances made in the techniquesand instrumentations in research in biological sciences are called as Bio- technology. Some people define bio- technology as - the manipulation of organisms to make products that benefit human beings.

The characteristics of a living organism are determined by some physical units called as genes that are arranged on chromosomes found inside the nucleus of a cell. A gene is made of molecules of deoxyribonucleic acid (DNA), which is called as the Genetic Blue Print of each living cell. The characteristics and structure of DNA was studied by Rosalind Franklin, Francis Crick, James Watson and Maurice Wilkins (1953), for which Crick, Watson and Wilkins were awarded Nobel Prize for the year 1962, though the contribution of Rosalind Franklin remained un-noticed and she died four years earlier to the award of the Nobel Prize to these workers. Later, Dr. H.G. Khorana, R.H. Holley and M.W. Nirenberg were awarded Nobel Prize for the year 1968, for their pioneering works related to the synthesis of gene.

These studies of gene and DNA formed the basis if development of bio- technological studies in India and abroad. The presentation of double helical model of DNA and discoveries pertaining to gene, mark the beginning of the modern era of Bio- technology.

Fig. Helical Structure of DNA Molecule

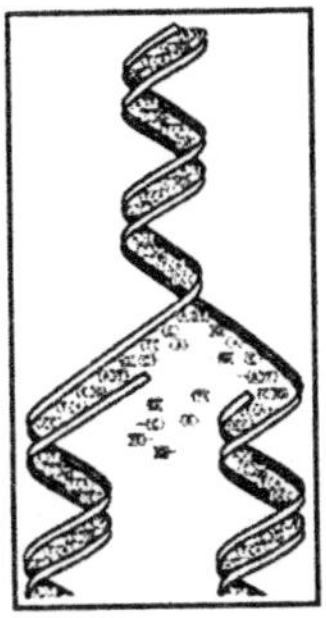

Fig. Double Helix of DNA

Origin and Development

The pioneering achievement in the field of Bio- technology was the production of greater genetic varieties of plants and animals through the techniques of Hybridization in 5000BC. It was through hybridization that present day High Yielding Varieties of plants and animals were evolved. This technique is being continuously used in improved forms in our programmes concerned with food production.

Norman E. Borlaug, the American Agriculturist, received the Nobel Prize for the year 1970 for his important contribution ofincreasing the world food supply through different techniques like hybridization, Selective Breeding and Gene Transfer.

Hybridization is the technology or process of crossing individuals of opposite sex of different species so as to produce off springs of combined character. The Indian Scientist Dr. M.S.Swaminathan initiated and developed technologies for increasing food production in India. Dr.Swaminathan is now being called as the Father of Economic Ecology.

In 1960, Werner Arber discovered specific enzymes called as Restriction Enzyme in bacteria. These enzymes can cut the DNA strand of an organism at a precise point. Further, Stanley Cohen and Herbert Boyer became successful in removing a specific gene from a bacterium and inserting it into another bacterium by the application of the restriction enzyme. This discovery led to the development of the Recombinant DNA Technology which is commonly called as Genetic Engineering.

Another major achievement in the field of bio-technological development was the manipulation of bacteria to produce Human Protein. Afterwards, the American Molecular Biologist Lydia Villa- Komaroff developed basic bio-technology for the production of Rat Insulin with the help of genetically engineered bacteria. By using genetically engineered bacteria, scientists became successful in producing Human Insulin in 1978 and within a period of 5 years the Human Insulin became the first biopharmaceutical product in the market to control Diabetes Mellitus.

Scope of Bio- Technology

Bio- technology has scope in the fields of Agriculture and Animal Husbandry, Molecular Medicine, Forensic Science, Microbial Genetics, Environmental Sciences etc.

Agriculture and Animal Husbandry

The scope of bio- technological applications in Agriculture and Animal Husbandry is concerned with- the development of disease and pest resistant varieties of plants and animals; the production of food containing high nutrient value; the production of vaccines and the development of drought resistant crop varieties.

Molecular Medicine

This field of Bio- technology incorporates- disease diagnosis through cultural and different biotechnological methods and through Gene Therapy (cure of diseases through improvement and modification of genes.

Forensic Science

This science relates to the identification of criminals and researches pertaining to crime. This field of bio- technology incorporates- the DNA- matching tests; the establishment of genetic relationships; DNA-mapping; DNA- Fingerprinting and Genetic Analysis etc.

Microbial Genetics

This field of bio- technology incorporates- Study of

microorganisms and their genetic make up; the study of genetic behaviour and possibilities of genetic alterations; recombination etc.The microbial Genetics is also concerned with the study of bio- weapons; bio-wars and bio- fuels.

Environmental Science

This field of bio- technology incorporates- studies pertaining to the detection of microorganisms causing health hazards; environmental pollutions and the monitoring of environmental problems concerned with the living world.

Applications of Biotechnology

The application of bio- technology can be categorized as Specific Applications and General Applications.

- *Genetic Engineering*: The experiments of Stanley Cohen and Herbert Boyer pertaining to the removal of specific gene from a bacterium and the insertion of the same gene into another bacterium, marks the beginning of the Recombinant DNA- technology which is also called as the Genetic Engineering. In their experiments, the"restriction enzymes"were used for cutting the segments of DNA. The Enzymes are specific proteins which can speed up biochemical processes without being altered in the chemical reactions.

The science of changing the pattern of development and behaviour of an organism by making alterations in its genes is called as genetic engineering. The organisms produced by altering their genetic make up through DNA recombinant technology or through genetic engineering are called as Genetically Modified Organisms or GMOs. In agriculture, GM- crops are also being produced through genetic engineering. The food produced fromgenetically modified crops is called as GM- food.

The genetically engineered organism in which a foreign gene has been introduced to obtain desired results is called as a Transgenic Organism. Such an organism may be a plant, an animal, bacteria or fungi. These organisms are developed by injecting a foreign gene into the embryo or the fertilized egg of an organism.

Importance of Genetic Engineering

- Through genetic engineering the alterations in the genetic make up of an organism can be done successfully,
- Genetic engineering can be employed in increasing food production from plants and animals,
- It is applied in the diagnosis of diseases and improvements in the medical treatment.
  - *Plant Tissue Culture*: The technique of growing plant tissues on artificial nutrient medium under laboratory conditions, so as to produce new plants, is called as plant tissue culture. These may be the Meristem Culture, Embryo Culture and Anther Culture and so on.
  - *Monoclonal Antibody*: The specialized types of protein molecules produced in the laboratory are called as monoclonal antibodies. These are produced naturally in our bodies when any bacteria or virus invades it. Monoclonal antibodies are produced in our blood and protect us from different types pf diseases. These bodies are used in identifying different types of cells. These are also employed in many diagnostic tests for bacteria and viruses. The experiments on using monoclonal antibodies for fighting against cancer are also going on across the world.

    The genetic engineers have engineered the bacterial species E. coli to synthesize a specific protein called as Interferon. These proteins are produced naturally by body cells to oppose viral infections. The biologically synthesized interferon has been tested and found successful.
  - *Synthesis of Enzymes*: The Proteinaceous chemical substances of biological origin that accelerate biochemical reactions without undergoing any

change are called as enzymes. The term "enzyme"was coined by William Kuhne in 1867 on the basis of his studies on yeast.

Specific enzymes can be synthesized through the application of bio- technology. These enzymes are used in various processes like removal of stains, softening of fabrics, preparation of digestible foods, processing of meat and, even the treatment of cancer.

- *Synthesis of Biodegradable Plastics*: Bio- technology is currently employed in the synthesis of plastic which is biodegradable i.e. unlike other plastics; this plastic can be broken down into simpler substances by microorganisms.

  The biodegradable plastic is made from lactic acid which is produced at the time of bacterial fermentation of plant materials like discarded stalks of corn. In the process, molecules of Lactic Acid are chemically grouped to form the biodegradable plastic. In fact, the biodegradable plastic is a material which has most of the properties of plastic except the property of being non- biodegradable.
- *Blood Substitute*: Today the number of people needing blood transfusion is increasing due to frequent accidents and diseases. In view of these facts, biotechnologists are trying to synthesize artificial blood through bio- technology. The biotechnologists like Mary L. Nucci and Abraham Abuchowski are likely to get success in their experiments in this regard.
- *Bio- technology and Mining*: Different types of bacteria are being currently employed in the extraction of different metals like copper, zinc, lead and other metals. These bacteria act on the metallic compounds available inside the earth and help in the isolation of respective metals.
- *Bio- technology and Blood Clotting Protein*: Bio-

technology has many types of dramatic applications in the field of medicine. It has produced Factor- VIII, which is a blood clotting protein and which is not produced in haemophilic persons. Under these conditions the haemophilic people are always at the risk of bleeding to death. In the process of synthesis of Factor VIII, the human genes having codes of production of blood clotting protein are transplanted into the haemophilic person. The factor VIII is being produced commercially since 1992.

- *Cloning*: The bio- technology of production of cells or organisms that are originally derived from a single original cell or organism by asexual method under laboratory conditions is called as cloning. The copies of organisms produced during cloning have identical genetic make up and are known as clones. We can define a clone as an individual cell or organism which has been grown by a single body cell and which is genetically identical to its parent cell.

  During 1950s, scientists cloned frogs and by 1980s they cloned mice. During 1996, Ian Wilmot and his team of researchers achieved success for the first time in cloning an adult sheep. They named the clone as Dolly. Scientists at Texas A & M University in College Station produced the first cloned cat on Dec. 22, 2001 through the Nuclear Transfer Technique. They named the clone - cc or the carbon copy. It is hoped that scientists may one day become successful in cloning extinct animals also.
- *Applications in food and beverage industry*: A number of food products are produced on industrial scale through the fermentation technology. Some of those products are wine, idlis, yoghurt, cheese, mycoproteins, quorn,

bread etc. Besides this bio- technology helps in the production of different vitamins, amino acids and vinegar etc.

- *Application of Bio- technology in Pharmaceuticals*: Modification of microorganisms, animals and plants so as to get maximum yield of medicinally useful substances is called as pharmaceutical bio-technology.

  Some important examples are being given below:

  a. Production of human insulin from non-human sources.
  b. Production of hormones like Interferon, Cytokinins, Steroids and human growth hormones.
  c. Gene-therapy for prevention and control of diseases.
  d. Development of vaccines and antibodies.

- *Application of bio- technology in Agriculture*: Tissue culture, cloning and hybridization are important bio- technologies that are being developed in agriculture today. Tissue culture is important for the propagation of high yielding varieties of plants for agriculture and floriculture. Hybridization is the technique of combining properties of two plants or animals to produce one better hybrid plant or animal. Though this process frequently occurs in nature, the adoption of this technology has supported the agriculture to produce more food to feed growing population in many countries. Similarly, hybridization technology has supported animal husbandry to produce more milk and meat.

  The application of bio- technology in agriculture can make it more sustainable. The introduction of bio-fertilizers in soil can improve it's composition besides making it fertile in a natural way. Similarly, the introduction of bio-pesticides can control pests through natural ways without contaminating the natural environment. The development of disease resistant and pest resistant crop varieties through bio- technological methods has further supported the agriculture. Scientists are continuously experimenting for the improvement of crop plants like potato, tomato, cabbage and other vegetables and fruits as well as other crop plants like sugarcane, wheat, maize etc. to obtain better yield. Some new crop varieties like Triticale(a man made cereal) have also been developed to add nutrients in our food and increase food production.

– *Application of bio- technology in Pollution Control*: The natural tendency of microorganisms can be exploited through bio- technology for solving the problem of wastes in the environment. The crop residues and animal waste are used in making manure by the activity of aerobic and anaerobic bacteria. The bacterial species Pseudomonas aeruginosa has been developed to eat away the oil spilled on the sea surface. Solid wastes like crop residues can be put inside a biogas digester to produce biogas, a most suitable source of rural energy.

In nature, green plants control atmospheric carbon dioxide by utilizing it in photosynthesis. Hence, plantation of more and more fast growing trees may be an important bio- technological method of controlling carbon dioxide level of atmosphere. Plants fix carbon dioxide by the help of specific enzyme Ribulose- bi- phosphate Carboxilase. The function of this enzyme is controlled genetically. Hence, scientists are trying to manipulate this enzyme to increase the rate of photosynthesis for greater reduction of atmospheric carbon dioxide level.

– *Application of bio- technology in Waste Water Treatment*: The treatment of waste water comprises three major steps- the Primary Treatment, the Secondary Treatment and the Tertiary Treatment. The primary treatment comprises many sub- steps like sedimentation, chemical coagulation and precipitation. These sub- steps remove most of the physical impurities or pollutants. The secondary treatment comprises biological process involving bio- technology of employing bacteria, fungi, algae etc. for the breaking down of complex pollutants. In this process, the effluent is passed through a microbial slime layer. The microbes present in this layer break down the organic and nitrogenous waste liberating carbon dioxide and nitrogen dioxide.

- *Application of bio- technology in the degradation of pesticides*: Different species of bacteria and fungi tend to degrade pesticides. These microorganisms can be genetically manipulated to degrade more and more of them. Some species of bacteria like Pseudomonas sp., Flavobacterium, Azotobacter, E. coli and Acromobacter tend to degrade different pesticides. It has been reported that a mixture of Phenerocheate (a fungus) and enzyme peroxidase in suitable proportion can degrade DDT.
- *Bioremediation*: The application of biotic agents like microorganisms in the correction and recovery of environmental damage is called as bioremediation. The removal of oil spilled on sea water by the help of bacteria is one example of bioremediation.
- *Industrial Applications of bio- technology*: Biotechnology is currently being applied in many areas of industry like the production of stain remover, detergents, bread, biotech-polyester, vitamins; stone washed jeans, bleached paper etc. *Here are some examples*:

a. Detergents containing protease enzyme can remove stains of proteinaceous nature both on fabrics and lenses,
b. Detergents containing lipase enzyme can remove stains of oil and grease, and those containing amylase can remove starch grains stuck with fabrics.Now a day, polyesters are being synthesized from corn starch feed stock through the application of biotechnology. The enzyme cellulose is used for fading of jeans.

Chapter 4

Recombinant DNA Technology

SYNTHESIS OF HUMAN INSULIN

NATURE AND PURPOSE OF SYNTHESISING HUMAN INSULIN

Since Banting and Best discovered the hormone, insulin in 1921. diabetic patients, whose elevated sugar levels are due to impaired insulin production, have been treated with insulin derived from the pancreas glands of abattoir animals. The hormone, produced and secreted by the beta cells of the pancreas' islets of Langerhans, regulates the use and storage of food, particularly carbohydrates.

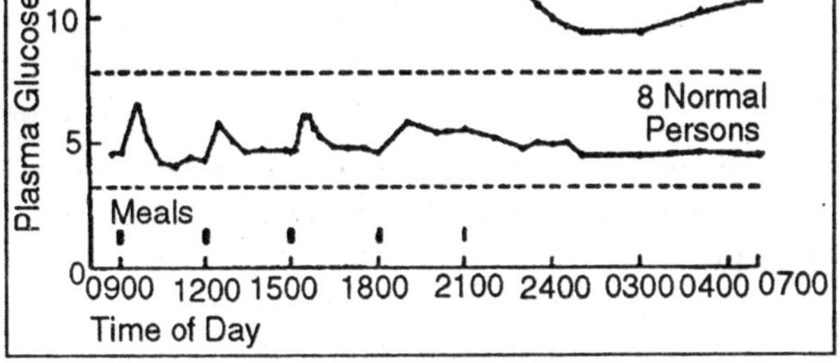

Although bovine and porcine insulin are similar to human insulin, their composition is slightly different. Consequently, a number of patients' immune systems produce antibodies against it, neutralising its actions and resulting in inflammatory responses at injection sites. Added to these adverse effects of bovine and porcine insulin, were fears of long term complications ensuing from the regular injection of a foreign substance, as well as a projected decline in the production of animal derived insulin. These factors led researchers to consider synthesising Humulin by inserting the insulin gene into a suitable vector, the E. coli bacterial cell, to produce an insulin that is chemically identical to its naturally produced counterpart. This has been achieved using Recombinant DNA technology. This method is a more reliable and sustainable method than extracting and purifying the abattoir by-product.

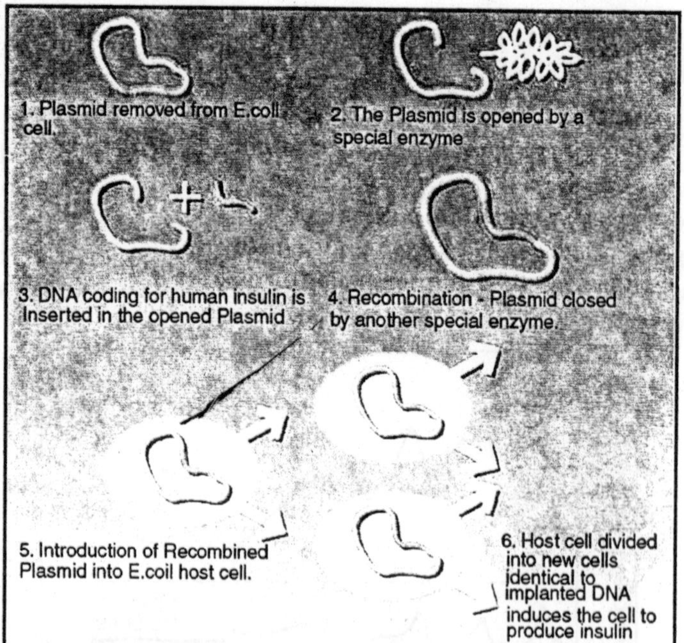

UNDERSTANDING THE GENETICS INVOLVED

The Structure of Insulin

Chemically, insulin is a small, simple protein. It consists

of 51 amino acid, 30 of which constitute one polypeptide chain, and 21 of which comprise a second chain. The two chains are linked by a disulfide bond.

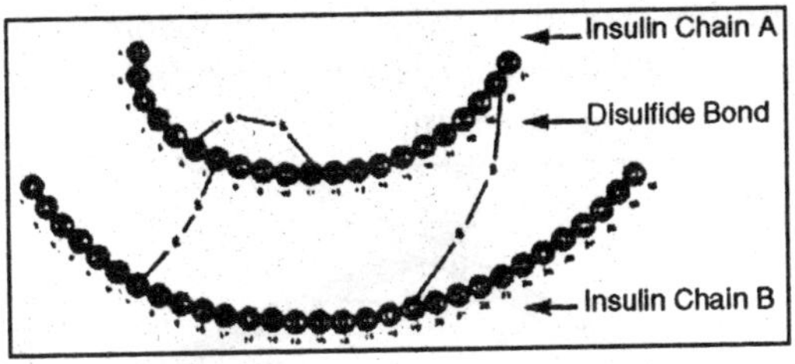

Inside the Double Helix

The genetic code for insulin is found in the DNA at the top of the short arm of the eleventh chromosome. It contains 153 nitrogen bases (63 in the A chain and 90 in the B chain).DNA Deoxyribolnucleic Acid), which makes up the chromosome, consists of two long intertwined helices, constructed from a chain of nucleotides, each composed of a sugar deoxyribose, a phosphate and nitrogen base.

There are four different nitrogen bases, adenine, thymine, cytosine and guanine. The synthesis of a particular protein such as insulin is determined by the sequence in which these bases are repeated.

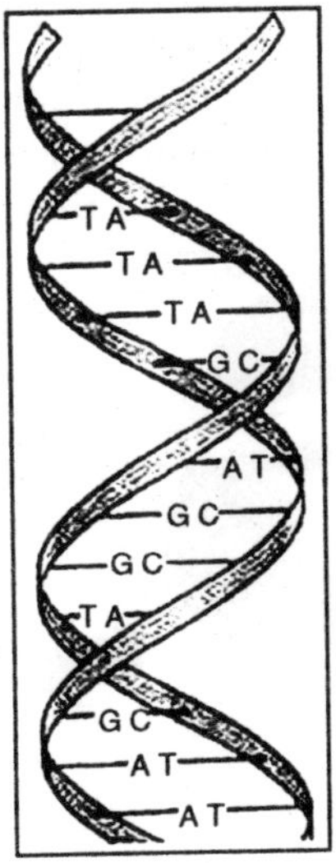

INSULIN SYNTHESIS FROM THE GENETIC CODE

The double strand of the eleventh chromosome of DNA divides in two, exposing unpaired nitrogen bases which are specific to insulin production.

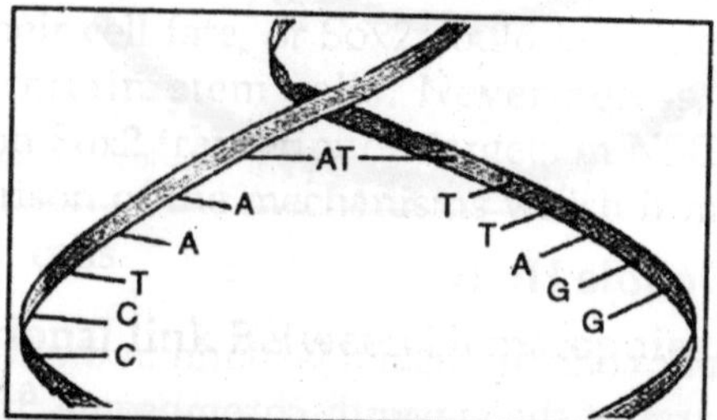

Using one of the exposed DNA strands as a template, messenger RNA forms in the process of transcription.

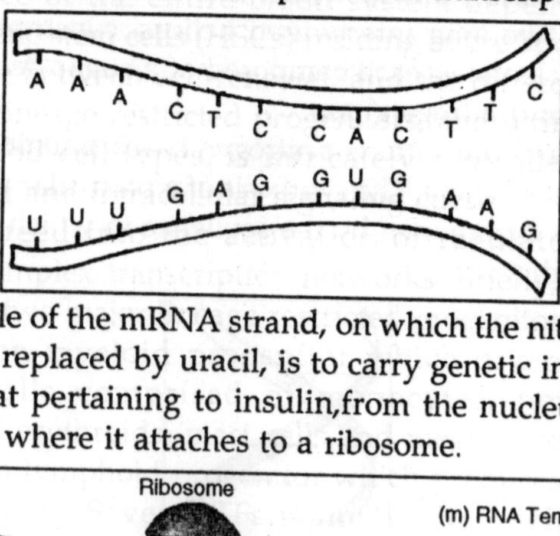

The role of the mRNA strand, on which the nitrogen base thymine is replaced by uracil, is to carry genetic information, such as that pertaining to insulin,from the nucleus into the cytoplasm, where it attaches to a ribosome.

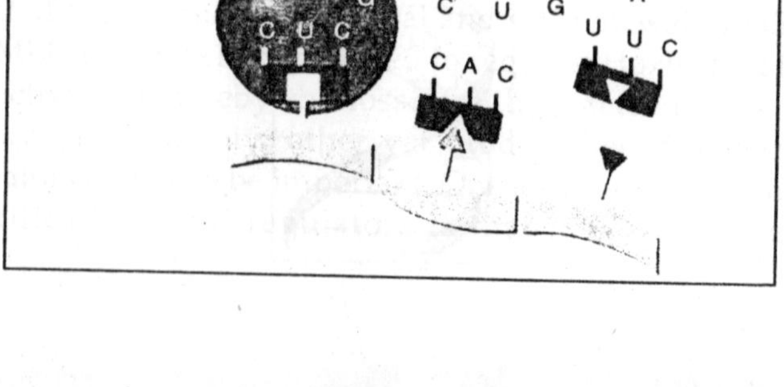

The nitrogen bases on the mRNA are grouped into threes, known as codons.

Transfer RNA (tRNA) molecules, three unpaired nitrogen bases bound to a specific amino acid, collectively known as an anti-codon pair with complementary bases (the codons) on the mRNA.

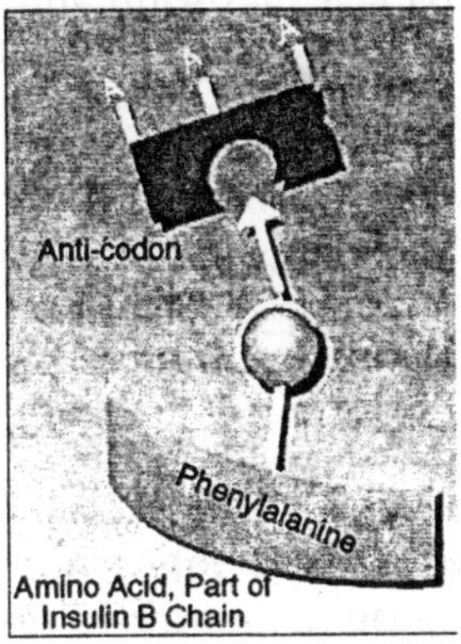

The reading of the mRNA by the tRNA at the ribosome is known as translation. A specific chain of amino acids is formed by the tRNA following the code determined by the mRNA. The base sequence of the mRNA has been translated into an amino acid sequence which link together to form specific proteins such as insulin.

THE VECTOR (GRAM NEGATIVE E. COLI)

A weakened strain of the common bacterium, Escherrichia coli (E. coli), an inhabitant of the human digestive tract, is the 'factory' used in the genetic engineering of insulin.

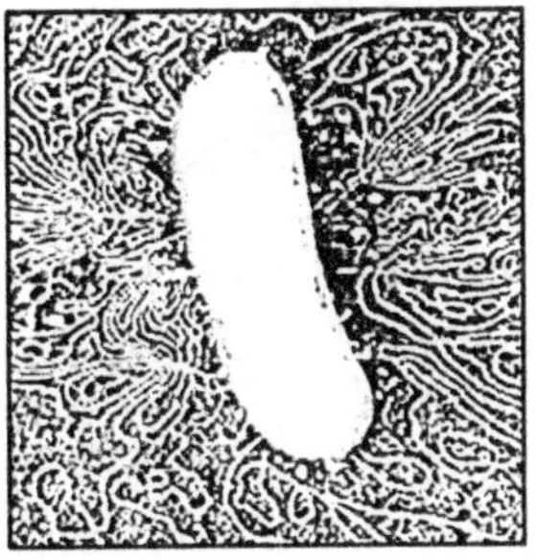

When the bacterium reproduces, the insulin gene is

replicated along with the plasmid, a circular section of DNA. E. coli produces enzymes that rapidly degrade foreign proteins such as insulin. By using mutant strains that lack these enzymes, the problem is avoided.

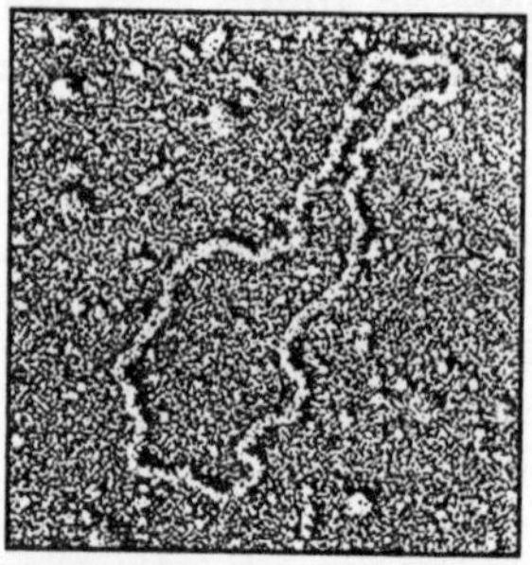

In E. coli, B-galactosidase is the enzyme that controls the transcription of the genes. To make the bacteria produce insulin, the insulin gene needs to be tied to this enzyme.

INSIDE THE GENETIC ENGINEER'S TOOLBOX

Restriction enzymes, naturally produced by bacteria, act like biological scalpels, only recognising particular stretches of nucleotides, such as the one that codes for insulin.

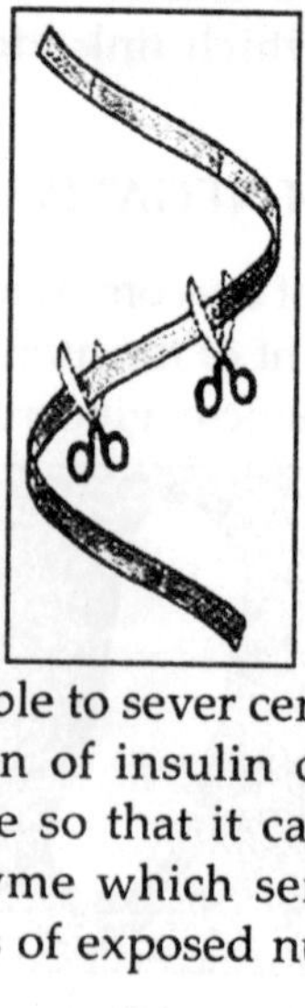

This makes it possible to sever certain nitrogen base pairs and remove the section of insulin coding DNA from one organism's chromosome so that it can manufacture insulin. DNA ligase is an enzyme which serves as a genetic glue, welding the sticky ends of exposed nucleotides together.

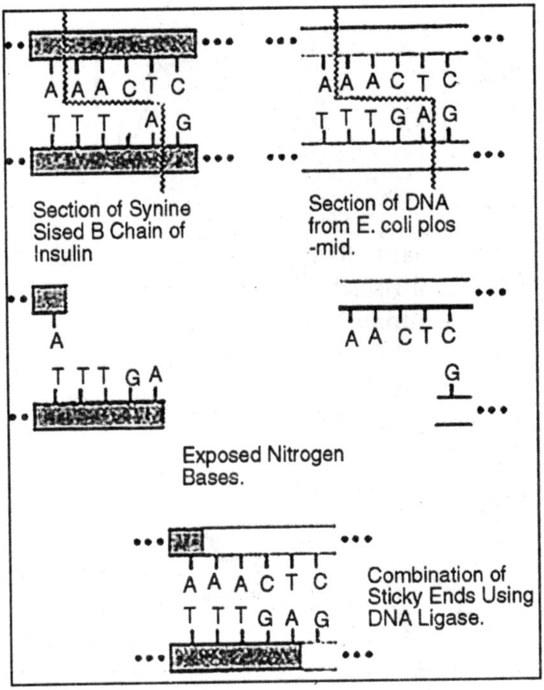

MANUFACTURING HUMULIN

The first step is to chemically synthesise the DNA chains that carry the specific nucleotide sequences characterising the A and B polypeptide chains of insulin.

The required DNA sequence can be determined because the amino acid compositions of both chains have been charted. Sixty three nucleotides are required for synthesising the A chain and ninety for the B chain, plus a codon at the end of each chain,signalling the termination of protein synthesis. An anti-codon, incorporating the amino acid, methionine, is then placed at the beginning of each chain which allows the removal of the insulin protein from the bacterial cell's amino acids. The synthetic A and B chain 'genes' are then separately inserted into the gene for a bacterial enzyme, B-galactosidase, which is carried in the vector's plasmid. At this stage, it is crucial to ensure that the codons of the synthetic gene are compatible with those of the B-galactosidase.

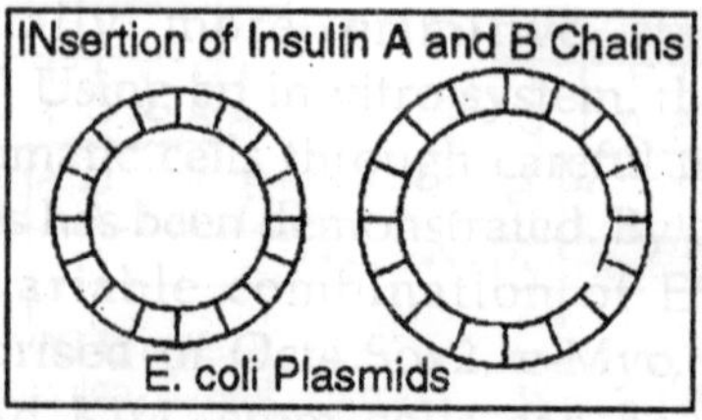

The recombinant plasmids are then introduced into E. coli cells. Practical use of Recombinant DNA technology in the synthesis of human insulin requires millions of copies of the bacteria whose plasmid has been combined with the insulin gene in order to yield insulin. The insulin gene is expressed as it replicates with the B-galactosidase in the cell undergoing mitosis.

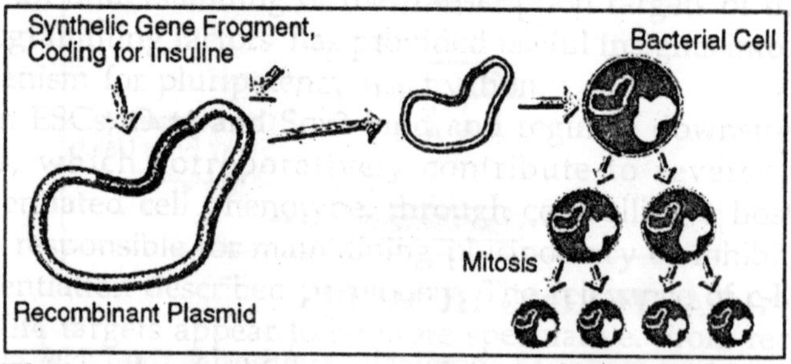

The protein which is formed, consists partly of B-galactosidase, joined to either the A or B chain of insulin. The A and B chains are then extracted from the B-galactosidase fragment and purified.

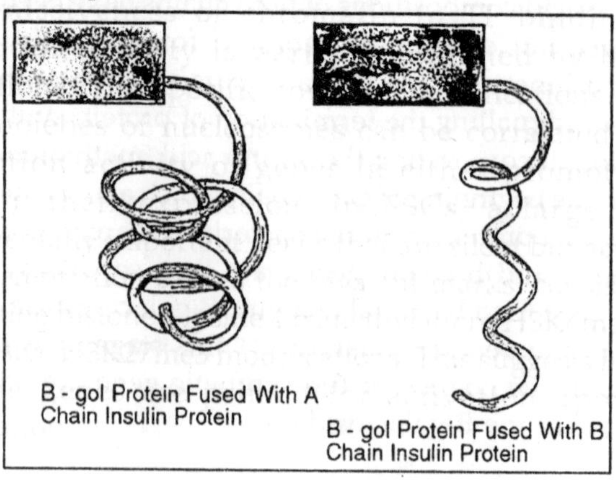

The two chains are mixed and reconnected in a reaction that forms the disulfide cross bridges, resulting in pureHumulin - synthetic human insulin.

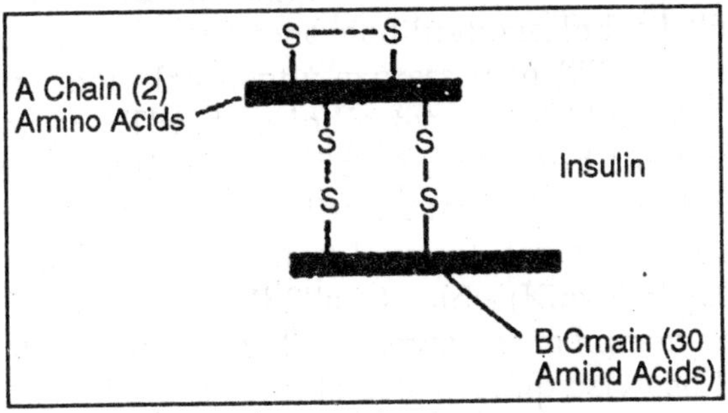

BIOLOGICAL IMPLICATIONS

Human insulin is the only animal protein to have been made in bacteria in such a way that its structure is absolutely identical to that of the natural molecule. This reduces the possibility of complications resulting from antibody production.

In chemical and pharmacological studies, commercially available Recombinant DNA human insulin has proven indistinguishable from pancreatic human insulin. Initially the major difficulty encountered was the contamination of the final product by the host cells, increasing the risk of contamination in the fermentation broth. This danger was eradicated by the introduction of purification processes.

When the final insulin product is subjected to a battery of tests, including the finest radio-immuno assay techniques, no impurities can be detected.

The entire procedure is now performed using yeast cells as a growth medium, as they secrete an almost complete human insulin molecule with perfect three dimensional structure. This minimises the need for complex and costly purification procedures.

HYPOGLYCAEMIC COMPLICATIONS

Since porcine insulin was phased out, and the majority of insulin dependent patients are now treated with genetically engineered recombinant human insulin, doctors and patients have become concerned about the increase in the number of hypoglycaemic episodes experienced.

Although hypoglycaemia can be expected occasionally with any type of insulin, some people with diabetes claim that they are less cognisant of attacks of hypoglycaemia since switching from animal derived insulin to Recombinant DNA human insulin.

In a British study, published in the 'Lancet", hypoglycaemia was induced in patients using either pork or human insulin, The researchers found"no significant difference in the frequency of signs of hypoglycaemia between users of the two different types of insulin."

An anecdotal report from a British patient who had been insulin dependent for thirty years, stated that she began experiencing recurring, unheralded hypoglycaemia only after substituting Recombinant DNA human insulin for animal derived insulin. After switching back to pork insulin to ease her mind, she hadn't experienced any unannounced hypoglycaemia. Eli Lilly and Co., a manufacturer of human insulin, noted that a third of people with diabetes, who have been insulin dependent for over ten years,"lose their hypoglycaemic warning signals, regardless of the type of insulin they are taking."

Dr Simon P. Wolff of the University College of London said in an issue of Nature ,"As far as I can make out, there's no fault (with the human insulin)."He concluded,"I do think we need to have a study to examine the possible risk."

Although the production of human insulin is unarguable welcomed by the majority of insulin dependent patients, the existence of a minority of diabetics who are unhappy with the product cannot be ignored. Although not a new drug, the insulin derived from this new method of production must continue to be studied and evaluated, to ensure that all its users have the opportunity to enjoy a complication free existence.

DNA CLONING

In biology a clone is a group of individual cells or

organisms descended from one progenitor. This means that the members of a clone are genetically identical, because cell replication produces identical daughter cells each time. The use of the word clone has been extended to recombinant DNA technology, which has provided scientists with the ability to produce many copies of a single fragment of DNA, such as a gene, creating identical copies that constitute a DNA clone. In practice the procedure is carried out by inserting a DNA fragment into a small DNA molecule and then allowing this molecule to replicate inside a simple living cell such as a bacterium.

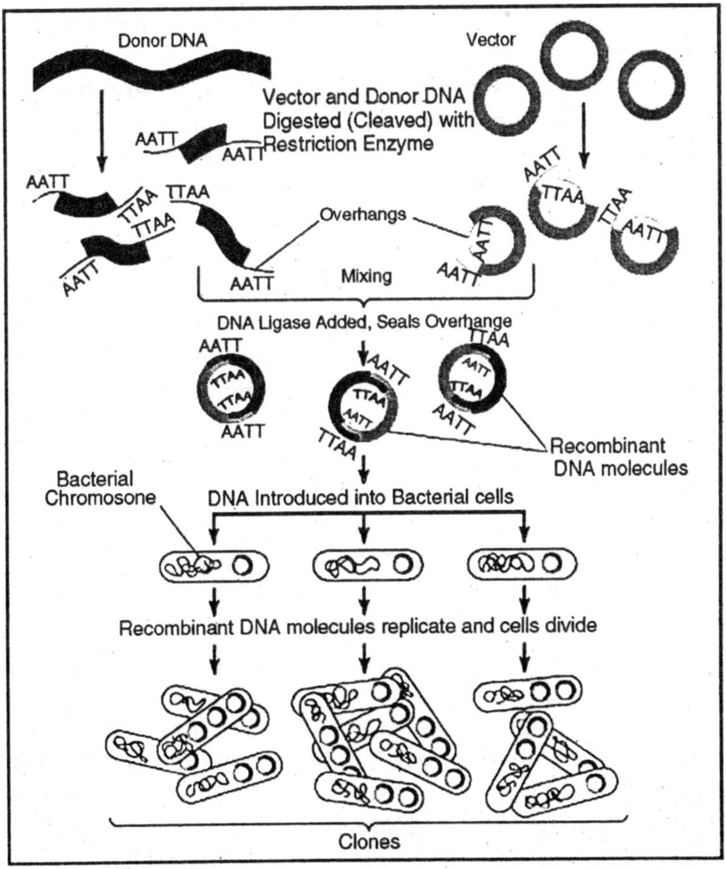

The small replicating molecule is called a DNA vector

(carrier). The most commonly used vectors are plasmids (circular DNA molecules that originated from bacteria), viruses, and yeast cells.

Plasmids are not a part of the main cellular genome, but they can carry genes that provide the host cell with useful properties, such as drug resistance, mating ability, and toxin production. They are small enough to be conveniently manipulated experimentally, and, furthermore, they will carry extra DNA that is spliced into them.

CREATING THE CLONE

The steps in cloning are as follows. DNA is extracted from the organism under study and is cut into small fragments of a size suitable for cloning. Most often this is achieved by cleaving the DNA with a restriction enzyme. Restriction enzymes are extracted from several different species and strains of bacteria, in which they act as defence mechanisms against viruses. They can be thought of as"molecular scissors,"cutting the DNA at specific target sequences.

The most useful restriction enzymes make staggered cuts; that is, they leave a single-stranded overhang at the site of cleavage.

These overhangs are very useful in cloning because the unpaired nucleotides will pair with other overhangs made using the same restriction enzyme. So, if the donor DNA and the vector DNA are both cut with the same enzyme, there is a strong possibility that the donor fragments and the cut vector will splice together because of the complementary overhangs. The resulting molecule is called recombinant DNA.

It is recombinant in the sense that it is composed of DNA from two different sources. Thus, it is a type of DNA that would be impossible naturally and is an artifact created by DNA technology.

The next step in the cloning process is to cut the vector with the same restriction enzyme used to cut the donor DNA. Vectors have target sites for many different restriction enzymes, but the most convenient ones are those that occur only once in the vector molecule. This is because the restriction enzyme then merely opens up the vector ring, creating a space for the insertion of the donor DNA segment. Cut vector DNA and donor DNA are mixed in a test tube, and the complementary ends of both types of DNA unite randomly. Of course, several types of unions are possible: donor fragment to donor fragment, vector fragment to vector fragment, and, most important, vector fragment to donor fragment, which can be selected for.

Recombinant DNA associations form spontaneously in the above manner, but these associations are not stable because, although the ends are paired, the sugar-phosphate backbone of the DNA has not been sealed. This is accomplished by the application of an enzyme called DNA ligase, which seals the two segments, forming a continuous and stable double helix.

The mixture should now contain a population of vectors each containing a different donor insert. This solution is mixed with live bacterial cells that have been specially treated to make their cells more permeable to DNA.

Recombinant molecules enter living cells in a process called transformation. Usually, only a single recombinant molecule will enter any individual bacterial cell. Once inside, the recombinant DNA molecule replicates like any other plasmid DNA molecule, and many copies are subsequently produced. Furthermore, when the bacterial cell divides, all of the daughter cells receive the recombinant plasmid, which again replicates in each daughter cell.

The original mixture of transformed bacterial cells is spread out on the surface of a growth medium in a flat dish (Petri dish) so that the cells are separated from one another. These individual cells are invisible to the naked eye, but as each cell undergoes successive rounds of cell division, visible colonies form.

Each colony is a cell clone, but it is also a DNA clone because the recombinant vector has now been amplified by replication during every round of cell division. Thus, the Petri dish, which may contain many hundreds of distinct colonies, represents a large number of clones of different DNA fragments.

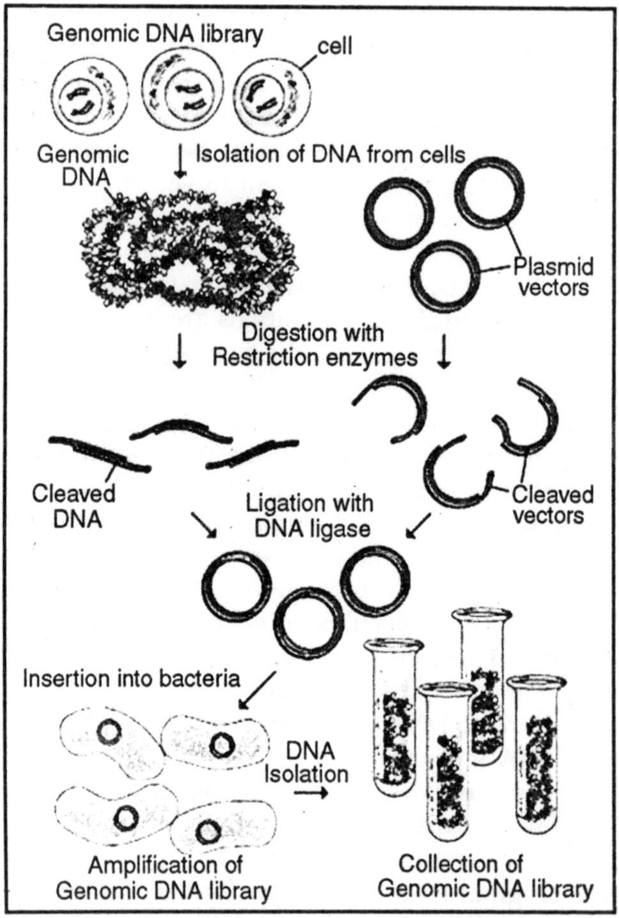

This collection of clones is called a DNA library. By considering the size of the donor genome and the average size of the inserts in the recombinant DNA molecule, a researcher can calculate the number of clones needed to encompass the entire donor genome, or, in other words, the number of clones needed to constitute a genomic library.

Another type of library is a cDNA library. Creation of a cDNA library begins with messenger ribonucleic acid(mRNA) instead of DNA. Messenger RNA carries encoded information from DNA to ribosomes for translation into protein. To create a cDNA library, these mRNA molecules are treated with the enzyme reverse transcriptase, which is used to make a DNA copy of an mRNA.

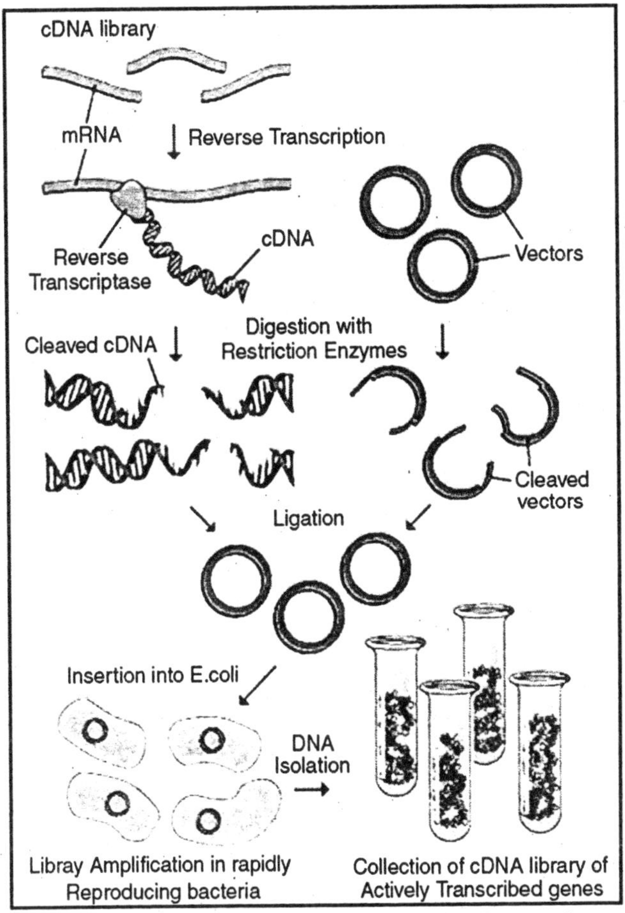

The resulting DNA molecules are called complementary DNA (cDNA). A cDNA library represents a sampling of the transcribed genes, whereas a genomic library includes untranscribed regions.

Both genomic and cDNA libraries are made without regard to obtaining functional cloned donor fragments. Genomic clones do not necessarily contain full-length copies of genes.

Furthermore, genomic DNA from eukaryotes (cells or

organisms that have a nucleus) contains introns, which are regions of DNA that are not translated into protein and cannot be processed by bacterial cells. This means that even full-sized genes are not translated in their entirety. In addition, eukaryotic regulatory signals are different from those used by prokaryotes (cells or organisms lacking internal membranes-i.e., bacteria).

However, it is possible to produce expression libraries by slicing cDNA inserts immediately adjacent to a bacterial promoter region on the vector; in these expression libraries, eukaryotic proteins are made in bacterial cells, which allows several important technological applications that are discussed below in DNA sequencing.

Several bacterial viruses have also been used as vectors. The most commonly used is the lambda phage. The central part of the lambda genome is not essential for the virus to replicate in Escherichia coli, so this can be excised using an appropriate restriction enzyme, and inserts from donor DNA can be spliced into the gap. In fact, when the phage repackages DNA into its protein capsule, it includes only DNA fragments the same length of the normal phage genome.

Vectors are chosen depending on the total amount of DNA that must be included in a library. Cosmids are engineered vectors that are hybrids of plasmid and phage lambda; however, they can carry larger inserts than either pUC plasmids (plasmids engineered to produce a very high number of DNA copies but that can accommodate only small inserts) or lambda phage alone.

Bacterial artificial chromosomes (BACs) are vectors based on F-factor (fertility factor) plasmids of E. coli and can carry much larger amounts of DNA. Yeast artificial chromosomes (YACs) are vectors based on autonomously replicating plasmids of Saccharomyces cerevisiae (baker's yeast). In yeast (a eukaryotic organism) a YAC behaves like a yeast chromosome and segregates properly into daughter cells. These vectors can carry the largest inserts of all and are used extensively in cloning large genomes such as the human genome.

ISOLATING THE CLONE

In general, cloning is undertaken in order to obtain the clone of one particular gene or DNA sequence of interest. The next step after cloning, therefore, is to find and isolate that clone among other members of the library. If the library encompasses the whole genome of an organism, then somewhere within that library will be the desired clone. There are several ways of finding it, depending on the specific gene concerned. Most commonly, a cloned DNA segment that shows homology to the sought gene is used as a probe. For example, if a mouse gene has already been cloned, then that clone can be used to find the equivalent human clone from a human genomic library. Bacterial colonies constituting a library are grown in a collection of Petri dishes.

Then a porous membrane is laid over the surface of each plate, and cells adhere to the membrane. The cells are ruptured, and DNA is separated into single strands-all on the membrane. The probe is also separated into single strands and labeled, often with radioactive phosphorus. A solution of the radioactive probe is then used to bathe the membrane. The single-stranded probe DNA will adhere only to the DNA of the clone that contains the equivalent gene. The membrane is dried and placed against a sheet of radiation-sensitive film, and somewhere on the films a black spot will appear, announcing the presence and location of the desired clone. The clone can then be retrieved from the original Petri dishes.

DNA SEQUENCING

Once a segment of DNA has been cloned, its nucleotide sequence can be determined. The nucleotide sequence is the most fundamental level of knowledge of a gene or genome. It is the blueprint that contains the instructions for building an organism, and no understanding of genetic function or evolution could be complete without obtaining this information.

Uses

Knowledge of the sequence of a DNA segment has many

uses, and some examples follow. First, it can be used to find genes, segments of DNA that code for a specific protein or phenotype.

If a region of DNA has been sequenced, it can be screened for characteristic features of genes. For example, open reading frames (ORFs)-long sequences that begin with a start codon (three adjacent nucleotides; the sequence of a codon dictates amino acid production) and are uninterrupted by stop codons (except for one at their termination)-suggest a protein-coding region.

Also, human genes are generally adjacent to so-called CpG islands-clusters of cytosine and guanine, two of the nucleotides that make up DNA. If a gene with a known phenotype (such as a disease gene in humans) is known to be in the chromosomal region sequenced, then unassigned genes in the region will become candidates for that function.

Second, homologous DNA sequences of different organisms can be compared in order to plot evolutionary relationships both within and between species.

Third, a gene sequence can be screened for functional regions. In order to determine the function of a gene, various domains can be identified that are common to proteins of similar function.

For example, certain amino acid sequences within a gene are always found in proteins that span a cell membrane; such amino acid stretches are called transmembrane domains. If a transmembrane domain is found in a gene of unknown function, it suggests that the encoded protein is located in the cellular membrane.

Other domains characterize DNA-binding proteins. Several public databases of DNA sequences are available for analysis by any interested individual.

Methods: The two basic sequencing approaches are the Maxam-Gilbert method, discovered by and named for American molecular biologists Allan M. Maxam and Walter Gilbert, and the Sanger method, discovered by English biochemist Frederick Sanger. In the most commonly used method, the Sanger method, DNA chains are synthesized on a template strand, but chain growth is stopped when one of four possible dideoxy nucleotides, which lack a 3′ hydroxyl group, is incorporated, thereby preventing the addition of another nucleotide.

A population of nested, truncated DNA molecules results that represents each of the sites of that particular nucleotide in the template DNA. These molecules are separated in a procedure called electrophoresis, and the inferred nucleotide sequence is deduced using a computer.

IN VITRO MUTAGENESIS

Another use of cloned DNA is in vitro mutagenesis in which a mutation is produced in a segment of cloned DNA. The DNA is then inserted into a cell or organism, and the effects of the mutation are studied. Mutations are useful to geneticists in enabling them to investigate the components of any biological process.

However, traditional mutational analysis relied on the occurrence of random spontaneous mutations-a hit-or-miss method in which it was impossible to predict the precise type or position of the mutations obtained. In vitro mutagenesis, however, allows specific mutations to be tailored for type and for position within the gene. A cloned gene is treated in the test tube (in vitro) to obtain the specific mutation desired, and then this fragment is reintroduced into the living cell, where it replaces the resident gene.

One method of in vitro mutagenesis is oligonucleotide-directed mutagenesis. A specific point in a sequenced gene is pinpointed for mutation. An oligonucleotide, a short stretch of synthetic DNA of the desired sequence, is made chemically. For example, the oligonucleotide might have adenine in one specific location instead of guanine.

This oligonucleotide is hybridized to the complementary strand of the cloned gene; it will hybridize despite the one base pair mismatch.

Various enzymes are added to allow the oligonucleotide to prime the synthesis of a complete strand within the vector. When the vector is introduced into a bacterial cell and replicates, the mutated strand will act as a template for a complementary strand that will also be mutant, and thus a fully mutant molecule is obtained.

This fully mutant cloned molecule is then reintroduced into the donor organism, and the mutant DNA replaces the resident gene.

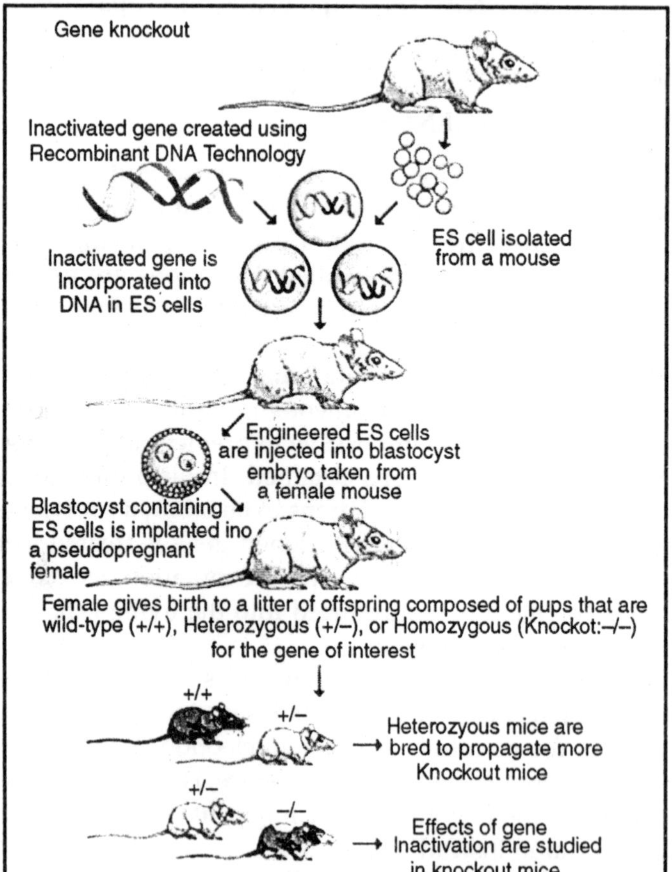

Another version of in vitro mutagenesis is gene disruption, or gene knockout. Here, the resident functional gene is replaced by a completely nonfunctional copy. The advantage of this technique over random mutagenesis is that specific genes can be knocked out at will, leaving all other genes untouched by the mutagenic procedure.

GENETICALLY MODIFIED ORGANISMS

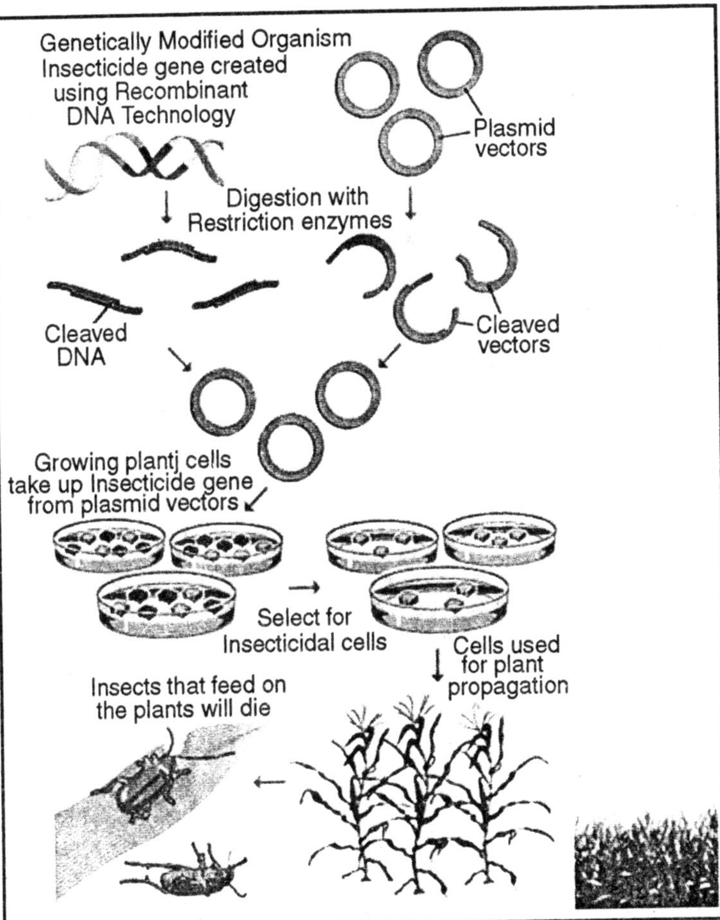

The ability to obtain specific DNA clones using recombinant DNA technology has made it possible to add the DNA of one organism to the genome of another. The added gene is called a transgene. The transgene inserts itself into a chromosome and is passed to the progeny as a new component of the genome. The resulting organism carrying the transgene is called a transgenic organism or a genetically modified organism (GMO). In this way, a"designer organism"is made that contains some specific change required for an experiment in basic genetics or for improvement of some commercial strain. Several transgenic plants have been produced. Genes for toxins that kill insects have been introduced in several species, including corn and cotton. Bacterial genes that confer resistance to herbicides also have been introduced into crop plants. Other plant transgenes aim at improving the nutritional value of the plant.

GENE THERAPY

Gene therapy is the introduction of a normal gene into an individual's genome in order to repair a mutation that causes a genetic disease. When a normal gene is inserted into a mutant nucleus, it most likely will integrate into a chromosomal site different from the defective allele; although this may repair the mutation, a new mutation may result if the normal gene integrates into another functional gene. If the normal gene replaces the mutant allele, there is a chance that the transformed cells will proliferate and produce enough normal gene product for the entire body to be restored to the undiseased phenotype.

So far, human gene therapy has been attempted only on somatic (body) cells for diseases such as cancer and severe combined immunodeficiency syndrome (SCIDS). Somatic cells cured by gene therapy may reverse the symptoms of disease in the treated individual, but the modification is not passed on to the next generation. Germinal gene therapy aims to place corrected cells inside the germ line (e.g., cells of the ovary or testis). If this is achieved, these cells will undergo meiosis and provide a normal gametic contribution to the next generation. Germinal gene therapy has been achieved experimentally in animals but not in humans.

Reverse Genetics

Recombinant DNA technology has made possible a type of genetics called reverse genetics. Traditionally, genetic research starts with a mutant phenotype, and, by Mendelian crossing analysis, a researcher is able to attribute the phenotype to a specific gene. Reverse genetics travels in precisely the opposite direction. Researchers begin with a gene of unknown function and use molecular analysis to determine its phenotype.

One important tool in reverse genetics is gene knockout. By mutating the cloned gene of unknown function and using it to replace the resident copy or copies, the resultant mutant phenotype will show which biological function this gene normally controls.

DIAGNOSTICS

Recombinant DNA technology has led to powerful diagnostic procedures useful in both medicine and forensics. In medicine these diagnostic procedures are used in counseling prospective parents as to the likelihood of having a child with a particular disease, and they are also used in the prenatal prediction of genetic disease in the fetus. Researchers look for specific DNA fragments that are located in close proximity to the gene that causes the disease of concern.

These fragments, called restriction fragment length polymorphisms (RFLPs), often serve as effective"genetic markers."In forensics, DNA fragments called variable number tandem repeats (VNTRs), which are highly variable between individuals, are employed to produce what is called a"DNA fingerprint."A DNA fingerprint can be used to determine if blood or other body fluids left at the scene of a crime belongs to a suspect.

GENOMICS

The genetic analysis of entire genomes is called genomics. Such a broadscale analysis has been made possible by the development of recombinant DNA technology. In humans, knowledge of the entire genome sequence has facilitated searching for genes that produce hereditary diseases. It is also capable of revealing a set of proteins-produced at specific times, in specific tissues, or in specific diseases-that might be targets for therapeutic drugs. Genomics also allows the comparison of one genome with another, leading to insights into possible evolutionary relationships between organisms.

Genomics has two subdivisions: structural genomics and functional genomics. Structural genomics is based on the complete nucleotide sequence of a genome. Each member of a library of clones is physically manipulated by robots and sequenced by automatic sequencing machines, enabling a very high throughput of DNA. The resulting sequences are then assembled by a computer into a complete sequence for every chromosome. The complete DNA sequence is scanned by computer to find the positions of open reading frames (ORFs), or prospective genes. The sequences are then compared to the sequences of known genes from other organisms, and possible functions are assigned. Some ORFs remain unassigned, awaiting further research.

RECOMBINANT DNA AND GENE CLONING

Recombinant DNA is DNA that has been created artificially. DNA from two or more sources is incorporated into a single recombinant molecule.

MAKING RECOMBINANT DNA (RDNA)

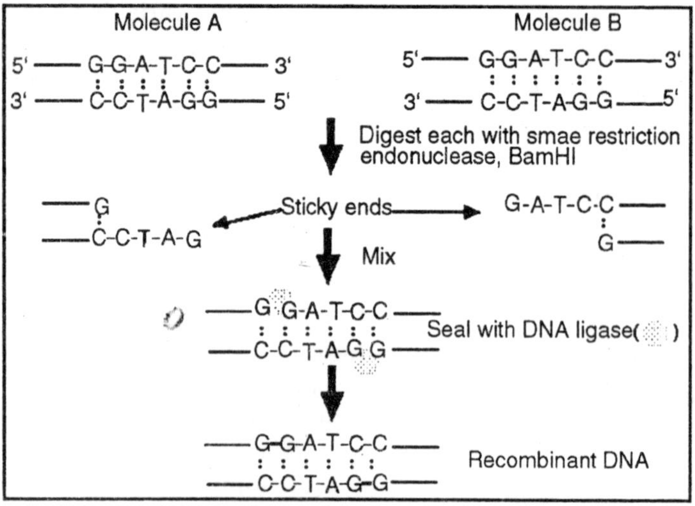

- Treat DNA from both sources with the same restriction endonuclease (BamHI in this case).

- BamHI cuts the same site on both molecules
  5' GGATCC 3'
  3' CCTAGG 5'
- The ends of the cut have an overhanging piece of single-stranded DNA.
- These are called"sticky ends"because they are able to base pair with any DNA molecule containing the complementary sticky end.
- In this case, both DNA preparations have complementary sticky ends and thus can pair with each other when mixed.
- DNA ligase covalently links the two into a molecule of recombinant DNA.

To be useful, the recombinant molecule must be replicated many times to provide material for analysis, sequencing, etc. Producing many identical copies of the same recombinant molecule is called cloning. Cloning can be done in vitro, by a process called the polymerase chain reaction (PCR). Here, however, we shall examine how cloning is done in vivo.

Cloning in vivo can be done in:

- Unicellular microbes like E. coli
- Unicellular eukaryotes like yeast and
- In mammalian cells grown in tissue culture.

In every case, the recombinant DNA must be taken up by the cell in a form in which it can be replicated and expressed. This is achieved by incorporating the DNA in a vector. A number of viruses (both bacterial and of mammalian cells) can serve as vectors. But here let us examine an example of cloning using E. coli as the host and a plasmid as the vector.

PLASMIDS

Plasmids are molecules of DNA that are found in bacteria separate from the bacterial chromosome.

They:

- Are small (a few thousand base pairs)
- Usually carry only one or a few genes
- Are circular
- Have a single origin of replication

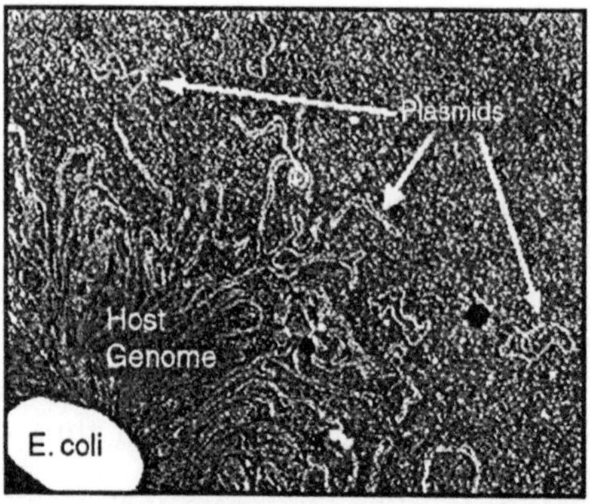

Electron micrograph of an E. coli cell ruptured to release its DNA. The tangle is a portion of a single DNA molecule containing over 4.6 million base pairs encoding approximately 4,300 genes. The small circlets are plasmids. Plasmids are replicated by the same machinery that replicates the bacterial chromosome.

Some plasmids are copied at about the same rate as the chromosome, so a single cell is apt to have only a single copy of the plasmid. Other plasmids are copied at a high rate and a single cell may have 50 or more of them. Genes on plasmids with high numbers of copies are usually expressed at high levels. In nature, these genes often encode proteins (e.g., enzymes) that protect the bacterium from one or more antibiotics. Plasmids enter the bacterial cell with relative ease. This occurs in nature and may account for the rapid spread of antibiotic resistance in hospitals and elsewhere. Plasmids can be deliberately introduced into bacteria in the laboratory transforming the cell with the incoming genes.

AN EXAMPLE

Courtesy of David Miklos and Greg Freyer of the Cold Spring Harbor Laboratory, who used these plasmids as the basis of a laboratory introduction to recombinant DNA technology that every serious biology student - high school or college - should experience!

pAMP:

- 4539 base pairs

- A single replication origin
- A gene (ampr)conferring resistance to the antibiotic ampicillin (a relative of penicillin)
- A single occurrence of the sequence
  5' GGATCC 3'
  3' CCTAGG 5'
  that, as we saw above, is cut by the restriction enzyme BamHI
- A single occurrence of the sequence
  5' AAGCTT 3'
  3' TTCGAA 5'
  that is cut by the restriction enzyme HindIII

Treatment of pAMP with a mixture of BamHI and HindIII produces:

- A fragment of 3755 base pairs carrying both the ampr gene and the replication origin
- A fragment of 784 base pairs
- Both fragments have sticky ends

pKAN

- 4207 base pairs
- A single replication origin
- A gene (kanr) conferring resistance to the antibiotic kanamycin.
- A single site cut by BamHI
- A single site cut by HindIII

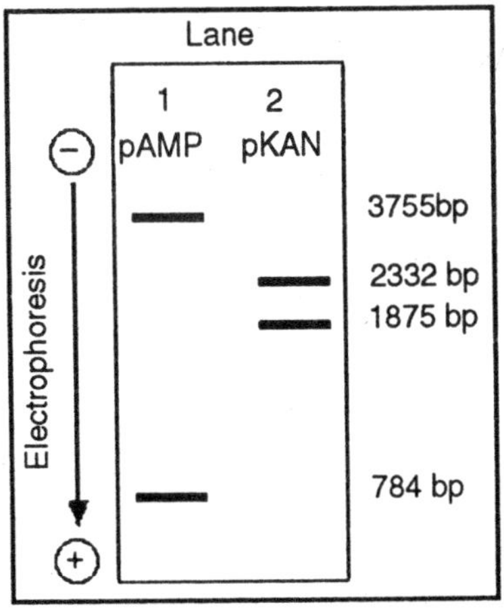

Treatment of pKAN with a mixture of BamHI and HindIII produces:

- A fragment of 2332 base pairs
- A fragment of 1875 base pairs with the kanr gene (but no origin of replication)
- Both fragments have sticky ends

These fragments can be visualized by subjecting the digestion mixtures to electrophoresis in an agarose gel. Because of its negatively-charged phosphate groups, DNA migrates towards the positive electrode (anode) when a direct current is applied. The smaller the fragment, the farther it migrates in the gel.

Ligation Possibilities

If you remove the two restriction enzymes and provide the conditions for DNA ligase to do its work, the pieces of these plasmids can rejoin (thanks to the complementarity of their sticky ends). Mixing the pKAN and pAMP fragments provides several (at least 10) possibilities of rejoined molecules. Some of these will not produce functional plasmids (molecules with two or with no replication origin cannot function).

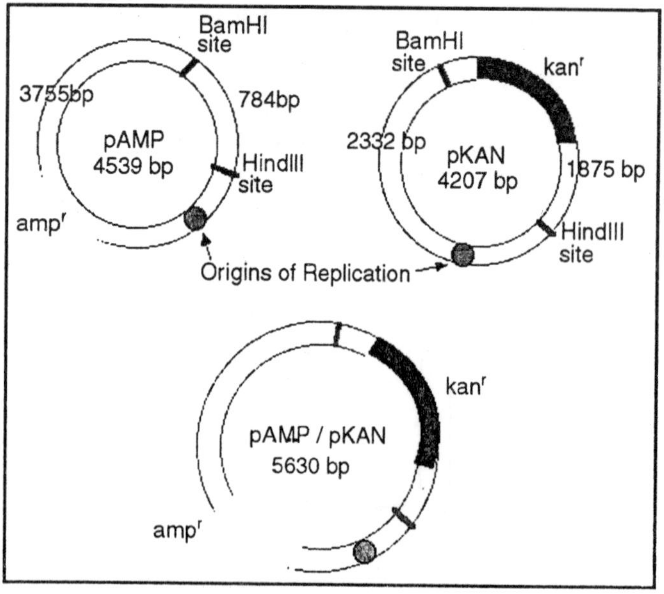

One interesting possibility is the joining of:

- The 3755-bp pAMP fragment (with ampr and a replication origin) with the
- 1875-bp pKAN fragment (with kanr)

Sealed with DNA ligase, these molecules are functioning plasmids that are capable of conferring resistance to both ampicillin and kanamycin. They are molecules ofrecombinant DNA. Because the replication origin, which enables the molecule to function as a plasmid, was contributed by pAMP, pAMP is called the vector. Transforming E. coli,

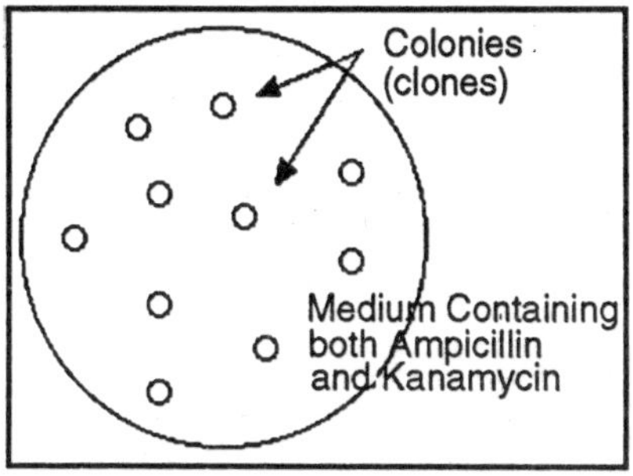

Treatment of E. coli with the mixture of religated molecules will produce some colonies that are able to grow in the presence of both ampicillin and kanamycin.

- A suspension of E. coli is treated with the mixture of religated DNA molecules.
- The suspension is spread on the surface of agar containing both ampicillin and kanamycin.
- The next day, a few cells - resistant to both antibiotics - will have grown into visible colonies containing billions of transformed cells.
- Each colony represents a clone of transformed cells.

However, E. coli can be simultaneously transformed by more than one plasmid, so we must demonstrate that the transformed cells have acquired the recombinant plasmid.

Electrophoresis of the DNA from doubly-resistant colonies (clones) tells the story:

- Plasmid DNA from cells that acquired their resistance from a recombinant plasmid only show only the 3755-bp and 1875-bp bands (Clone 1, lane 3).

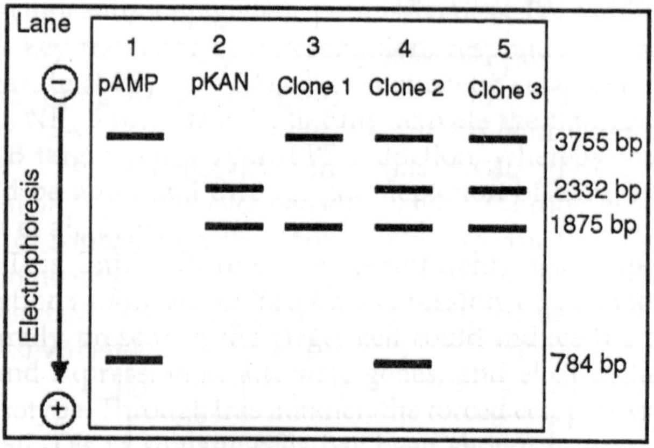

- Clone 2 (Lane 4) was simultaneous transformed by religated pAMP and pKAN. (We cannot tell if it took up the recombinant molecule as well.)
- Clone 3 (Lane 5) was transformed by the recombinant molecule as well as by an intact pKAN.

CLONING OTHER GENES

The recombinant vector described above could itself be a useful tool for cloning other genes. Let us assume that within its kanamycin resistance gene (kanr) there is a single occurrence of the sequence

5' GAATTC 3'

3' CTTAAG 5'

This is cut by the restriction enzyme EcoRI, producing sticky ends. If we treat any other sample of DNA, e.g., from human cells, with EcoRI, fragments with the same sticky ends will be formed.

Mixed with EcoRI-treated plasmid and DNA ligase, a small number of the human molecules will become incorporated into the plasmid which can then be used to transform E. coli. But how to detect those clones of E. coli that have been transformed by a plasmid carrying a piece of human DNA?

The key is that the EcoRI site is within the kanr gene, so when a piece of human DNA is inserted there, the gene's function is destroyed.

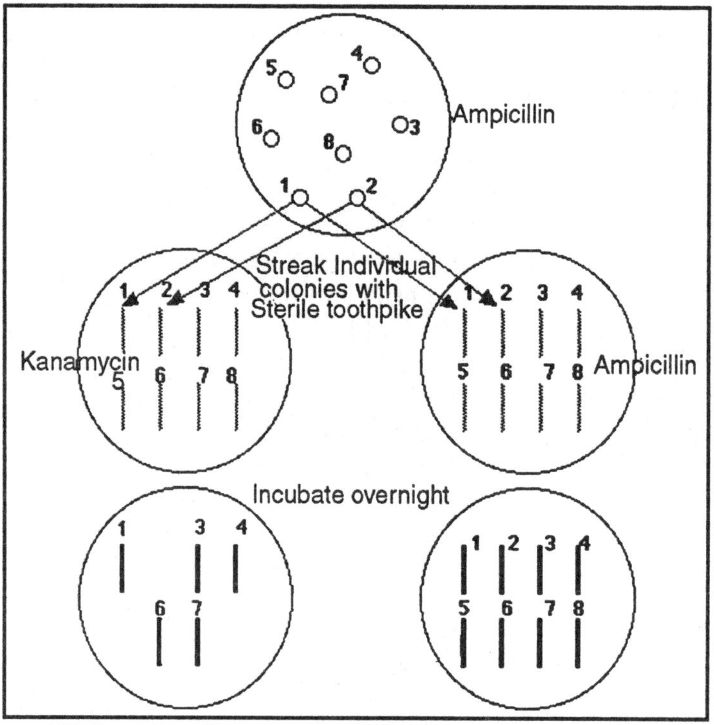

All E. coli cells transformed by the vector, whether it carries human DNA or not, can grow in the presence of ampicillin. But E. coli cells transformed by a plasmid carrying human DNA will be unable to grow in the presence of kanamycin.

So:

- Spread a suspension of treated E. coli on agar containing ampicillin only
- Grow overnight
- With a sterile toothpick transfer a small amount of each colony to an identified spot on agar containing kanamycin
- (do the same with another ampicillin plate)
- Incubate overnight

All those clones that continue to grow on ampicillin but fail to grow on kanamycin (here,clones 2, 5, and 8) have been transformed with a piece of human DNA.

DNA TECHNOLOGY

DEFINITION

Recombinant DNA (rDNA) technology is a field of molecular biology in which scientists"edit"DNA to form new synthetic molecules, which are often referred to as"chimeras". The practice of cutting, pasting, and copying DNA dates back to Arthur Kornberg's successful replication of viral DNA in a breakthrough that served as a proof-of-concept for cloning. This was followed by the Swiss biochemist Werner Arber's discovery of restriction enzymes in bacteria that degrade foreign viral DNA molecules while sparing their own DNA. Arber effectively showed geneticists how to"cut"DNA molecules; soon to follow was the understanding that ligase could be used to"glue"them together.

These two achievements launched rDNA technology research, allowing Man to"play god"for the first time in human history. It also touched off a firestorm among activists such as Jeremy Rifkin, for whom terms like 'cloning' and 'recombinant DNA' conjured image of rogue scientists running amok, creating Frankensteins and using them for malign purposes, or, worse yet, losing control of their own creations.

But these fears were quieted when scientists like Paul Berg, who won the Nobel Prize in 1980 for his groundbreaking work in rDNA research, halted potentially dangerous experiments so that public policy could deal with the quandaries posed by new technological possibilities. At the Asilomar Conference in 1975, the leading recombinant DNA specialists came together in Monterey , California , and agreed upon the need for safeguards to prevent any health crises or ecological disasters, and to reassure the public that their research would proceed with caution.

Other milestones in the development of recombinant DNA include the collaboration between Stanley Cohen and Herbert Boyer in 1972, and the 1976 founding of Genentech, the first company to work with rDNA in its drug development labs. In 1978 scientists were able to replicate Somatostatin, the protein that regulates human growth hormones. Since then, a host of other drugs have been developed through rDNA research, including Herceptin and Epogen.

So how does one go about making an rDNA molecule? There are three basic methods: Transformation, Phage Introduction, and Non-Bacterial Transformation. No matter which method is used, the goal is to introduce recombinant genes into a host cell along with expression factor, so that the host cell expresses the desired protein. In each case, scientists need to"turn off"the signals that tell a host cell to destroy or degrade the genes being introduced.

Recombinant DNA research is a challenging field, but it holds great promise for the future. In the future, rDNA technologies will play a key role in preventing genetic diseases, producing targeted medicines, and providing patients with less toxic pharmaceuticals. It will also impact agriculture and livestock as researchers find ways to optimize the genetic codes of plants and animals to resist disease. Even so, rDNA also poses serious questions.

Are we equipped to deal with the environmental impact of an rDNA molecule gone awry, e.g. genetically modified plants spreading beyond control and driving out local species? Is it a good thing for parents to 'pick and choose' the genetic traits they want in a child? And what effect will such decisions have on society as a whole?

WHAT IS GENE TECHNOLOGY

All gene manipulation is based on microbial genetics--ways of doing in the test tube what bacteria and viruses do naturally. Several of this week's gene manipulations are exemplified by the article on overexpressed angiopoietin in a transgenic mouse. Gel Electrophoresis Images and text based on MIT Hypertextbook.

This technique separates molecules on the basis of their size:

- Cast slab of gel material, usually agarose or polyacrylamide. The gel is a matrix of polymers forming sub-microscopic pores.
- The size of the pores can be controlled by varying the chemical composition of the gel.

The gel is set up for electrophoresis in a tank holding pH buffer.

Electrodes apply an electric field:

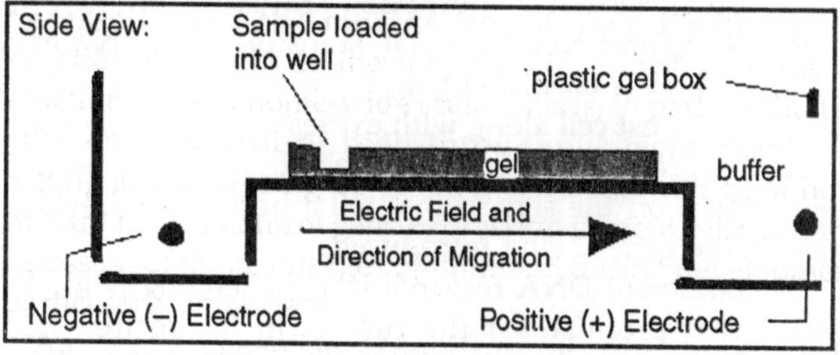

Fig. MIT Hypertextbook

The molecules to separate (DNA RNA) carry a net negative charge (why?) so they move along the electric field towards the positive cathode. (To separate proteins, a detergent would be included which coats the protein with negative charge.)

The larger molecules are held up as they try to pass through the pores of the gel, while the smaller molecules are impeded less and move faster. This results in separation by size, with the larger molecules nearer the well and the smaller molecules farther away.

Note that this separates on the basis of size (volume in solution), which is not necessarily molecular weight.

For example:

- Two DNA molecules of the same molecular weight will run differently if one is supercoiled, because the supercoils constrain the shape to be smaller.
- Two RNA molecules of the same molecular weight will run differently if one has much intramolecular base pairing, making it"smaller."

Aside from the above exceptions, the distance migrated is roughly proportional to the log of the inverse of the molecular weight (the log of 1/MW). Gels are normally depicted as running vertically, with the wells at the top and the direction of migration downwards. This leaves the large molecules at the top and the smaller molecules at the bottom.

Molecular weights are measured with different units for DNA, RNA, and protein:

- *DNA*: Molecular weight is measured in base-pairs, or bp, and commonly in kilobase-pairs (1000bp), or kbp.
- *RNA*: Molecular weight is measured in nucleotides, or nt, and commonly in kilonucleotides (1000nt), or knt. [Sometimes, bases, or b and kb are used.]
- *Protein*: Molecular weight is measured in Daltons (grams per mole), or Da, and commonly in kiloDaltons (1000Da), or kDa.

Molecular weight standards run in one well of the gel are used to calibrate the molecular weights of sample molecules. Below is a gel stained with a dye: a coloured molecule which binds to a specific class of macromolecules in a sequence-independent manner (probes bind in a sequence-dependent manner).

Sample 1: Contains only one size class of macromolecule - it could be a plasmid, a pure mRNA transcript, or a purified protein. In this case, you would not have to use a probe to detect the molecule of interest since there is only one type of molecule present. Blotting is usually necessary for samples that are not complex mixtures. By interpolation, its molecular weight is roughly 3.

Sample 2: Is what a sample of total DNA cut with a restriction enzyme, total cellular RNA, or total cellular protein would look like in a gel stained with a sequence-independent stain. There are so many bands that it is impossible to find the one we are interested in. Without a probe (which acts like a sequence-dependent stain) we cannot get very much information from a sample like this.

Different stains are used for different classes of macromolecules. DNA and RNA are generally stained withethidium bromide (EtBr), an intercalating agent. The

DNA-EtBr complex fluoresces under UV light. Protein is stained with Coomassie Blue or Silver Stain.

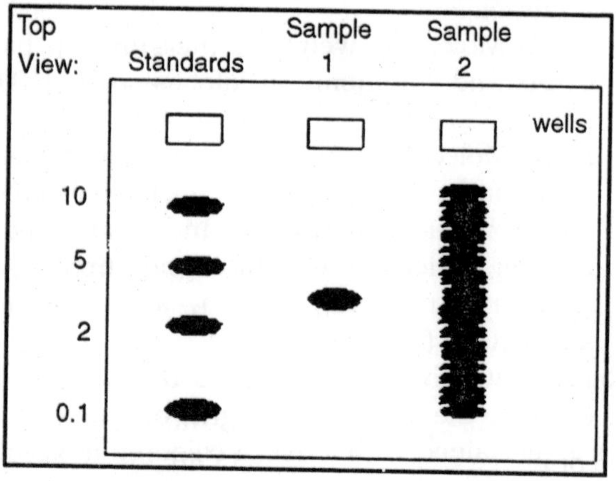

Fig. MIT Hypertextbook

CLONING GENES

In nature, DNA molecules recombine for various functions -- even DNA between different species. But twenty years ago, despite the work of Barbara McClintock and others, the extent of this recombination was not appreciated. DNA was still thought to be the"master molecule,"not to be violated by"unnatural"manipulation. When scientists began to manipulate DNA in the test tube, many scientists feared that disastrous monsters would result, with unspecified dangers to people.

In 1977 scientists at the Asilomar Conference proposed sweeping regulation on so-called"recombinant DNA, "technologies which recombine DNA from different species in the test tube.

Since then, the dangers have appeared to be little more than those of"natural"genetic mixing.

But we remain concerned about issues such as:

- Engineering food crops to resist pesticides. The pesticide resistance genes can escape into natural populations of weeds.
- Engineering a human symbiont microbe, such as E.

coli, to produce a deadly toxin such as botulin. In theory this could be done, although it's not clear where such an organism would live, or how well it could"compete"with natural flora.

- Societal dilemmas of human cloning. How far shall we use reproductive technology to shape future humans?

TECHNIQUES OF RECOMBINANT DNA

How do we manipulate these natural processes for biotechnology; for instance, to make a bacterium that produces large quantities of insulin?

One approach would be to cut the appropriate gene from human DNA and paste, or splice, it into a vector such as a plasmid or phage DNA. Our"scissors"are the class of enzymes called restriction endonucleases

RESTRICTION ENDONUCLEASES

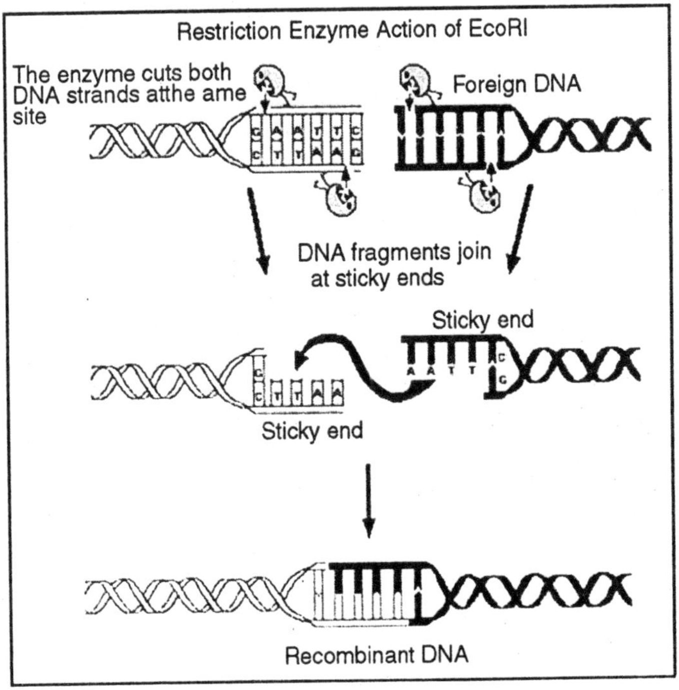

An"endonuclease"is an enzyme that cuts duplex DNA in the middle, not at an end (for exonuclease). Different species of bacteria have evolved different restriction endonucleases, each to cut foreign DNA that gets into their cells by mistake. To be cut, the DNA has to lack their own pattern of protective methylation. There are well over a hundred restriction enzymes, each cutting in a very precise way a specific base sequence of the DNA molecule. A restriction endonuclease cuts DNA only at a specific site, usually containing 4-6 base pairs. The enzyme has to cut the DNA backbone twice, recognizing the same type of site; therefore, the site"reads"the same way backwards as forwards--a palindrome. This"sticky ends"from two different DNA molecules can hybridize together; then the nicks are sealed using ligase. (Where does ligase come from? What is its natural function?) The result isrecombinant DNA. When this recombinant vector is inserted into E. coli, the cell will be able to process the instructions to assemble the amino acids for insulin production. More importantly, the new instructions are passed along to the next generation of E. coli cells in the process known as gene cloning.

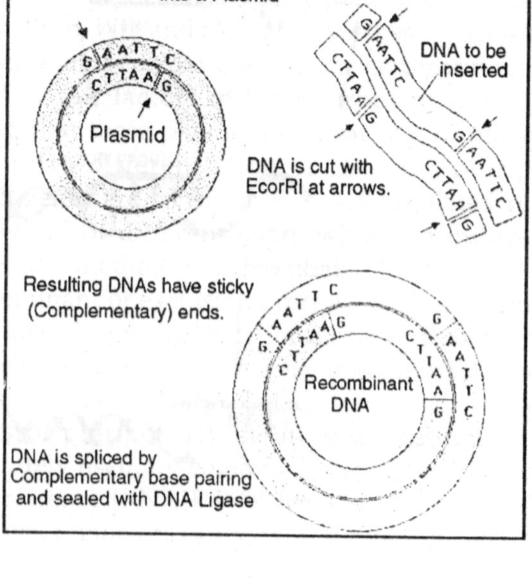

Restriction Site Analysis

How can we use restriction sites to analyse the plasmid products of ligation, and tell whether we in fact have ligated the correct molecule:

Problem: Suggest several"incorrect"ways the plasmid could recombine.

More Problems: Use the PLASMID Programme. You MUST practice restriction analysis with this programme; it will be on the quiz and/or the test.

How do we get the recombinant molecule into a bacterial cell? Usually by transformation (for a plasmid) or byin vitro packaging into a phage head coat (for a phage vector such as lambda phage).

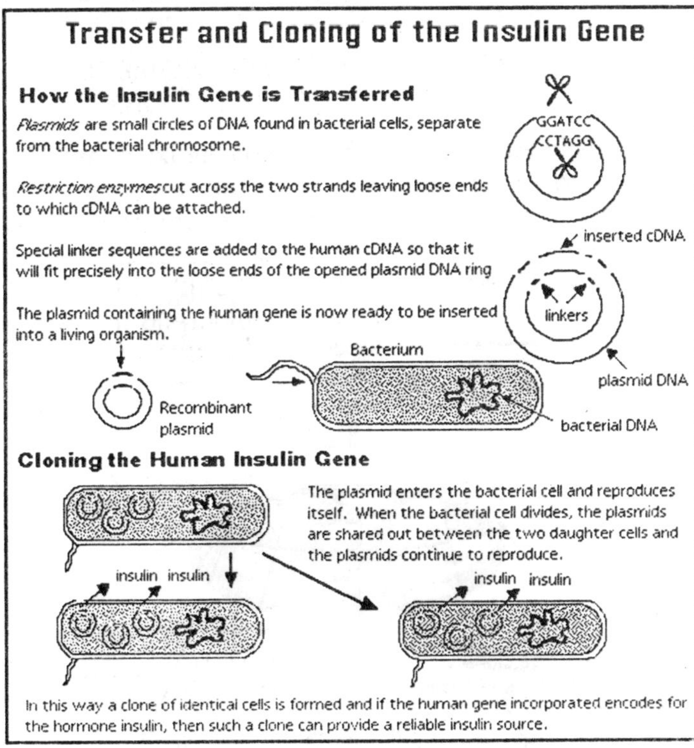

The above is a highly simplified description of recombinant DNA technology.

How would we actually locate the appropriately cloned gene? There are many different ways, depending on the specific case.

Here is one example, in which a partial sequence of the protein enables us to reverse the code and determine an approximate DNA sequence to use for a radiolabeled probe. The DNA probe willhybridize to clones containing the correct DNA, even if it is just one piece cut out of an entire genome.

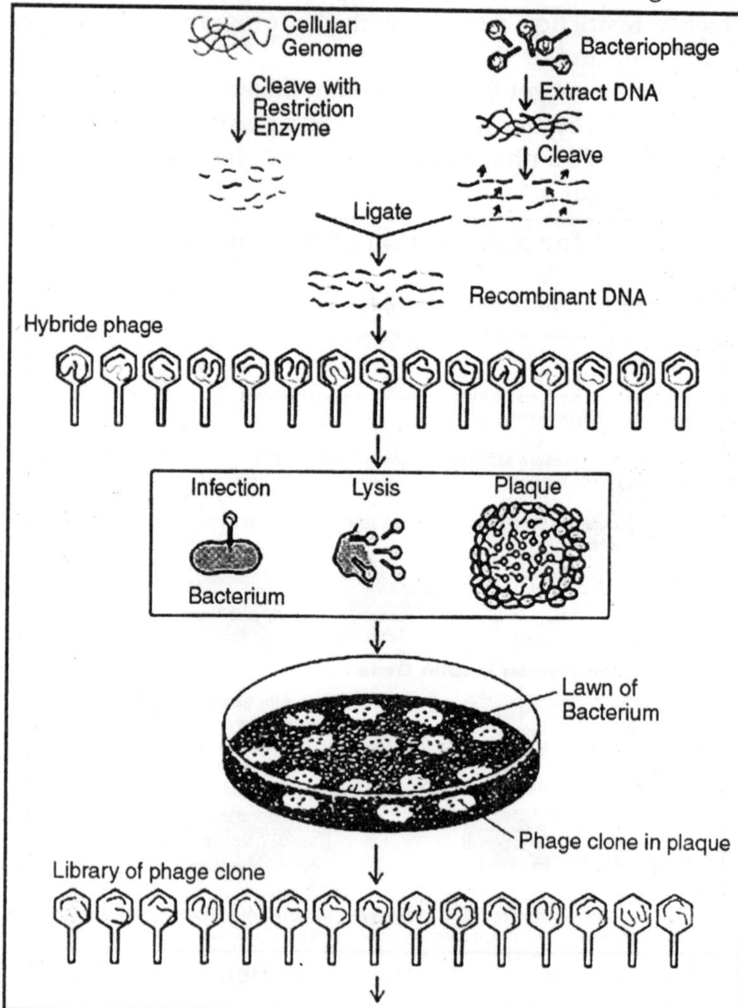

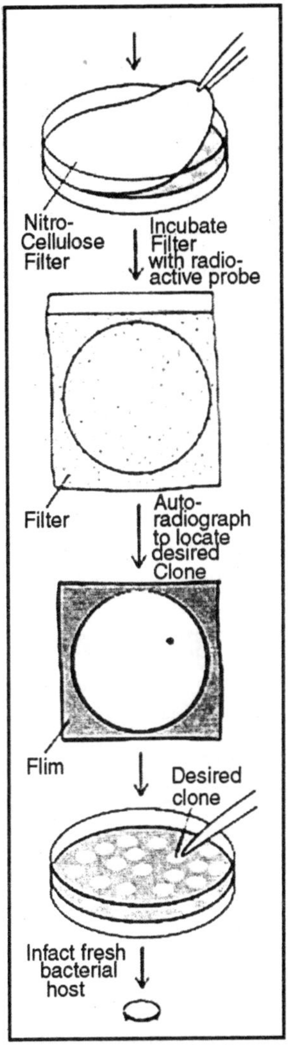

The radioactive probe is made by determining a short segment of the protein sequence, then"back translating"to the possible DNA sequences. Short DNA sequences are synthesized to match the protein sequence. Then these DNA oligomers (known as"oligos") are radiolabeled, and applied to the blotted clones. They should hybridize only to clones containing sequence encoding the desired protein.

Reverse transcription: cDNA Cloning

Suppose we need to clone a gene containing lots of introns. What will happen when the bacterium tries to express it? To overcome this problem, we can start with mRNA isolated from tissues that produce the desired protein. We then use reverse transcriptase enzyme (produced by a retrovirus related to HIV) to reverse transcribe the mRNA into a DNA molecule that now is free of introns. Now we can ligate"sticky ends"onto the cDNA and recombine it into a phage or plasmid vector.

Problem: Know the differences between genomic cloning and cDNA cloning. Explain the relative ADVANTAGES and DISADV-ANTAGES of each technique--depending on the aim of your research.

PCR

In PCR, a heat-stable DNA polymerase is used, most commonly Taq Polymerase from the thermophilic microbe Thermus aquaticus. Thomas Brock discovered T. aquaticus from a hot spring at Yellowstone National Park.

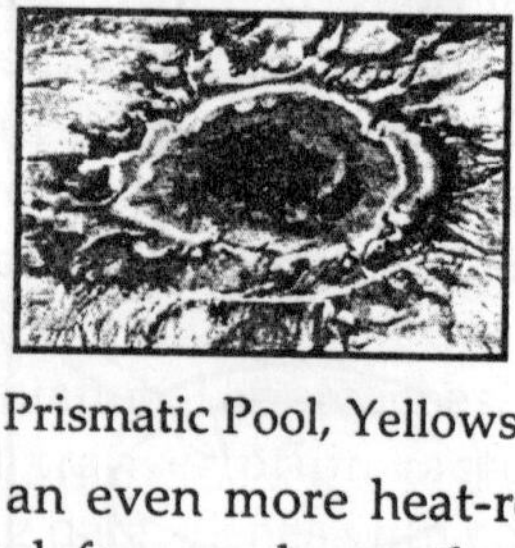

Fig. Prismatic Pool, Yellowstone

More recently, an even more heat-resistant polymerase has been developed from a hyperthermophilic microbe growing at 110 degrees C in hydrothermal vent ecosystems in the deep ocean; it's called"Vent Polymerase."

The Taq Polymerase is put with the DNA to be amplified, plus all four NTPs, plus two primers facing each other, about 200 - 6000 kb apart. The primers are selected based on the DNA region you want to amplify. The tube is placed in a thermal cycler. DNA gets synthesized from each primer, for about 2 minutes. Then the temperature is raised to 95 degrees C -- enough to denature (split apart) the DNA base pairs. But the Taq Polymerase remains intact, because it comes from an organism that evolved to grow at this temperature. Now the temperature is decreased again, and primers again can hybridize to the DNA--both the old AND the newly synthesized strands. Again, Taq Polymerase extends new DNA strands. Again, the temperature is raised.

After repeated cycles, the amount of DNA sequence between the two primers increases exponentially. First 2 strands, then 4, 8, 16, up to about a million. Thus, in a couple of hours, you can get million-fold amplification of a DNA sequence.

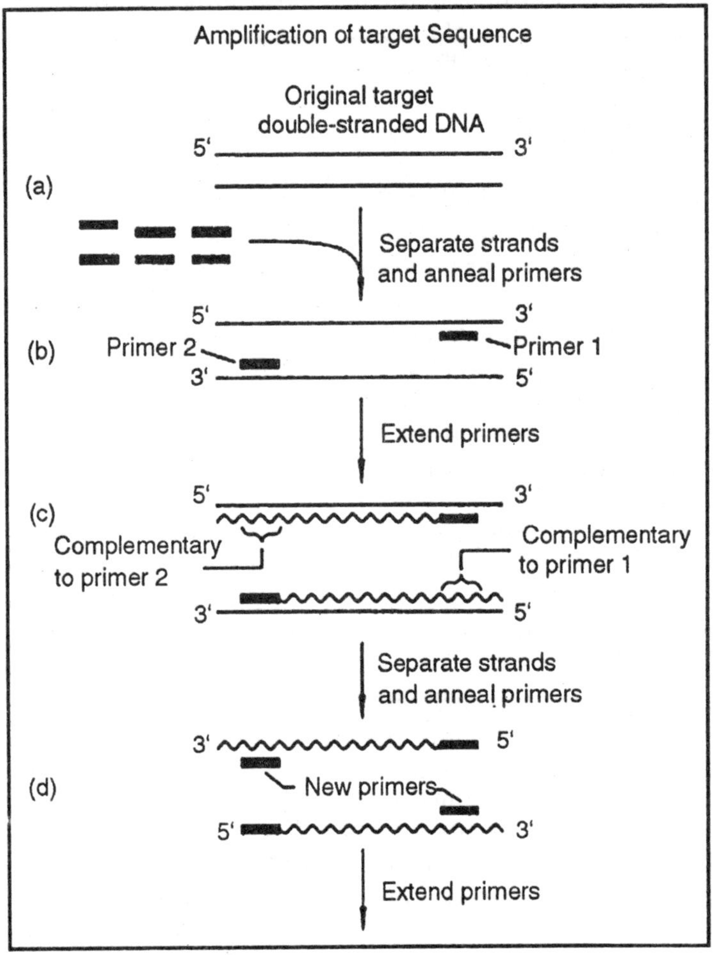

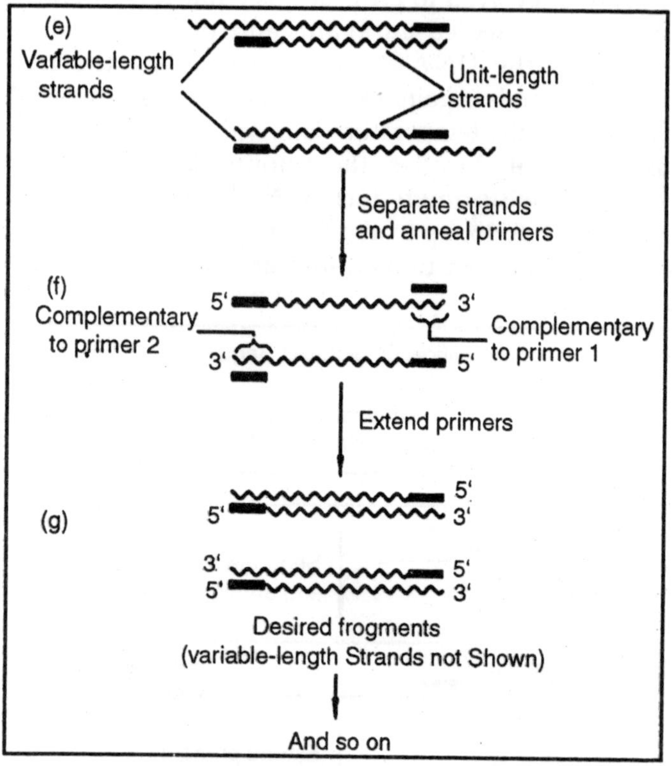

APPLICATIONS OF PCR

PCR has replaced cloning for many purposes, particularly the sequencing of DNA. It is faster and requires no vectors, which can mutate as they reproduce. It can be used forensically, to amplify tiny amounts of DNA from criminal evidence; or clinically, to detect DNA sequences linked to inherited disorders. Students characterized the microbial species of various environments in Knox County. Microbial colonies were placed directly into PCR reactions containing primers for amplification of ribosomal RNA genes (rDNA). Lane 4 contains an amplified band of the predicted size, 1000 bp. The top of each well contains genomic DNA. The smears at the bottom contain the PCR primers. The DNA will be sequenced and matched through GenBank to determine the microbial genus. Thanks to Dan Nickerson '00 and Adam Marks '01.

The Main Limitations of PCR

- Only relatively short sequences can be amplified reliably. Anything more than 10,000 base pairs is unlikely to be amplified.
- You need to know the right primer sequences to use, at both ends of the sequence you want to amplify. If two related genes have the same end sequences, you might amplify the wrong gene.
- You only obtain a DNA fragment. To see this DNA at work inside a living organism, some type of cloning has to be done.
- PCR Technology - discussion of preparation of the sample, the master mix, and primers, and detection and analysis of the reaction products.
  Gene sequence analysis Modified from MIT Hypertextbook.
- Restriction mapping. Construct a"road map"of restriction sites.
- DNA Sequence Analysis. Cut and clone various restriction fragments, and determine the exact sequence of base pairs. All sequence information is deposited in GenBank. If you just want the sequence of the peptide translated from the RNA, you have to look for insulin mRNA or cDNA.

Once we have a piece of DNA cloned, it is amplified and we now have a living clonewhich provides, in theory, an indefinite source of the DNA sequence. How do we analyse the sequence?

DNA SEQUENCE DETERMINATION

Once you have identified a particular region of DNA of interest, you need to find out the precise sequence of DNA nucleotides. This is done by di-deoxy sequencing, in which a DNA polymerase is put together with dNTPs in four different reactions, each containing a small amount of one di-deoxy NTP (ATP, TTP, CTP, or GTP).

The di-deoxy nucleotide lacks a 3'OH to continue chain extension, so the chain terminates. Each reaction produces a population of fragments terminated at A, T, C, or G. The fragments are either radiolabeled or enzymatically labeled. They can be separated on a gel, or on a fluorescence analyser. All published DNA sequences in the world are deposited in GenBank.

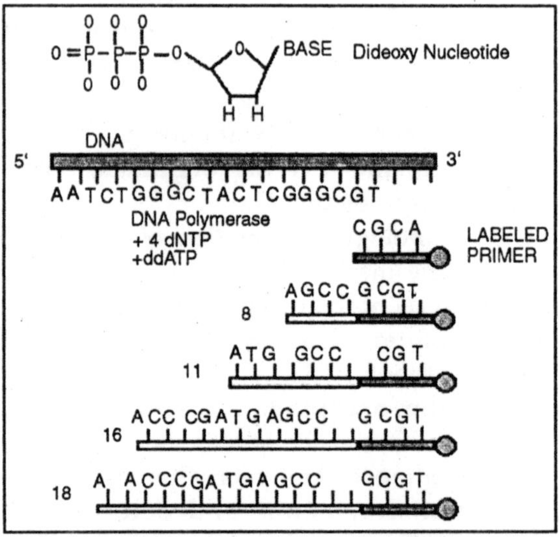

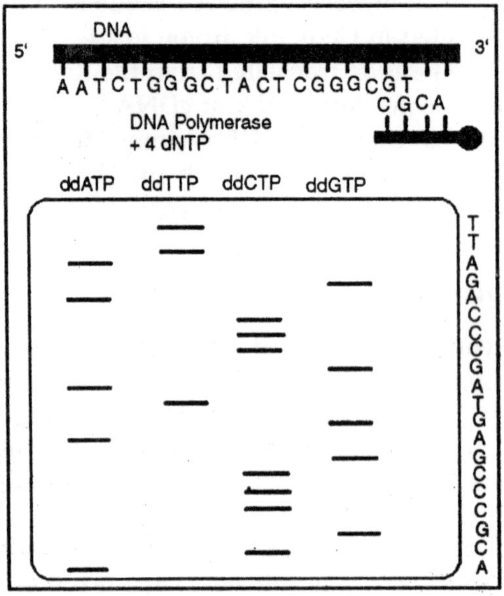

(c) Different-length strands can lined up by size determine DNA sequence.

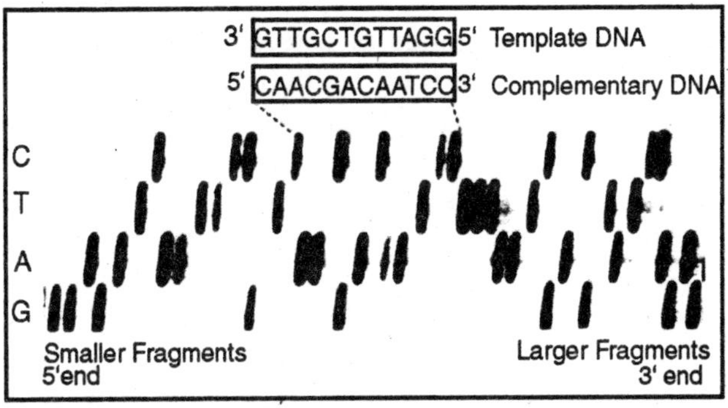

Historically, the products of dideoxy sequencing reactions were subjected to electrophoresis in four different lanes of a gel. Each lane contained a reaction using a different dideoxy terminator nucleotide. The data looked like the image at right.

Sequences of DNA in the gel lanes are read by a computerized fluorescence detection system that measures the intensity of light emission from each"band."The final output is a"trace"or"electrophoretogram"that plots the intensity of different colour emissions vs. the length of the DNA being sequenced. By observing the progression of peaks of different colours, the DNA sequence is derived (A C G T).

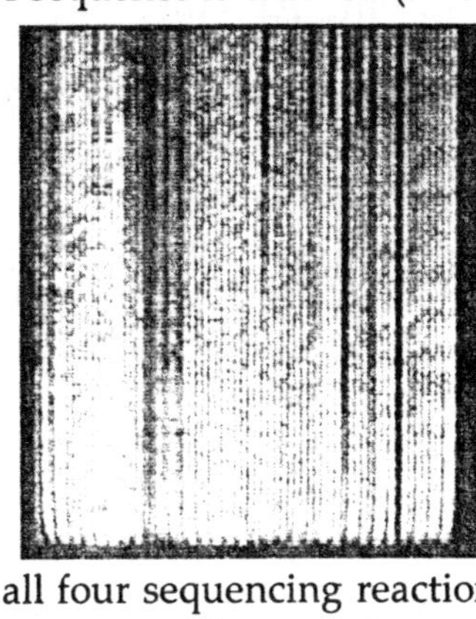

Products of all four sequencing reactions are loaded in a single gel lane or capillary tube and subjected to electrophoresis. Molecular labels consist of fluorescent dyes instead of radioactive nucleotides.

COMPLEMENTARITY AND HYBRIDIZATION

How do we find genes of interest in a genome? Or a particular mRNA in the total cellular RNA? Or a particular protein out of all cell protein? In solution, hybrid molecular complexes (usually called hybrids) of the following types can exist (other combinations are possible):

- DNA-DNA. A single-stranded DNA molecule (ssDNA probe) can form a double-stranded, base-paired hybrid with a ssDNA target if the probe sequence is the reverse complement of the target sequence. A radiolabeled DNA probe can be applied to DNA from a gel transferred to a membrane, called a Southern Blot.
- DNA-RNA. A single-stranded DNA (ssDNA) probe molecule can form a double-stranded, base-paired hybrid with an RNA (RNA is usually a single-strand) target if the probe sequence is the reverse complement of the target sequence. An RNA can be radiolabeled to probe a Southern Blot; or, a ssDNA probe can be applied to membrane-bound RNA, called a Northern Blot.
- Protein-Protein. An antibody probe molecule (antibodies are proteins) can form a complex with a target protein molecule if the antibody's antigen-binding site can bind to an epitope (small antigenic region) on the target protein. In this case, the hybrid is called an 'antigen-antibody complex' or 'complex' for short. A radiolabeled antibody can probe membrane-bound proteins, called a Western Blot (an even worse pun.)

There are two important features of hybridization:

- Hybridization reactions are specific - the probes will only bind to targets with complimentary sequence (or, in the case of antibodies, sites with the correct 3-d shape).
- Hybridization reactions will occur in the presence of large quantities of molecules similar but not identical

to the target. That is, a probe can find one molecule of target in a mixture of zillions of related but non-complementary molecules.

These properties allow you to use hybridization to perform a molecular search for one DNA molecule, or one RNA molecule, or one protein molecule in a complex mixture containing many similar molecules.

Southern, Northern, and Western Blots. Link to summary. Blots are named for the target molecule. Southern Blot--DNA cut with restriction enzymes - probed with radioactive DNA. Northern Blot--RNA - probed with radioactive DNA or RNA.

Example--used to measure angiopoietin angiopoietin expression from cDNA in transgenic mouse. Western Blot--Protein - probed with radioactive or enzymatically-tagged antibodies.These molecules must then be immobilized on a solid support, so that they will remain in position during probing and washing. The probe is then added, the non-specifically bound probe is removed, and the probe is detected. The place where the probe is detected corresponds to the location of the immobilized target molecule.

This process is diagrammed below:

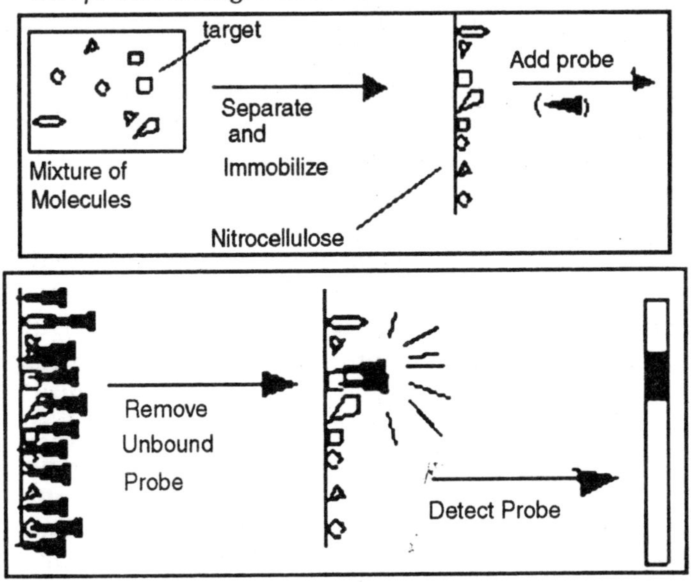

In the case of Southern, Northern, and Western blots, the initial separation of molecules is done on the basis of molecular weight, by gel electrophoresis.

Preparing for Blots

- Southern Blots. DNA is first cut with restriction enzymes and the resulting double-stranded DNA fragments have an extended rod conformation without pre-treatment.
- Northern Blots. Although RNA is single-stranded, RNA molecules often have small regions that can form base-paired secondary structures. To prevent this, the RNA is pre-treated with formaldehyde.
- Western Blots. Proteins have extensive 2' and 3' structures and are not always negatively charged. Proteins are treated with the detergent SDS (sodium dodecyl sulfate) which removes 2' and 3' structure and coats the protein with negative charges.

Transfer to Solid Support. After the DNA, RNA, or protein has been separated by molecular weight, it must be transferred to a solid support before hybridization. (Hybridization does not work well in a gel.) This transfer process is called blotting and is why these hybridization techniques are called blots. Usually, the solid support is a sheet of nitrocellulose paper (sometimes called a filter because the sheets of nitrocellulose were originally used as filter paper), although other materials are sometimes used. DNA, RNA, and protein stick well to nitrocellulose in a sequence-independent manner.

After a series of treatment steps, the probe is added. The probe hybridized to the target molecules is visualized either by autoradiography or by enzyme reaction.

Summary: The important properties of the three blotting procedures of DNA analysis:

	Southem	Northem	Westem
What is separated by molecular weight? (target)	DNA cut with restriction enzymes	RNA dena-tured with formaldehyde	Protein denatured with SDS

Probe	Radioactive gene X DNA	Radioactive gene X DNA	Antibody against pr-otein X, lab-eled with Radioactiv-or enzyme
What do you learn from it?	Restriction map of gene X in Chromo-some	-how much gene X mRNA is present? -how long is gene X mRNA?	-how much protein X is present? -how big is protein X?

DNA MICROARRAYS

We can now put most of the protein-encoding genes onto a microarray chip, using technology based on the DNA silicon chip industry. The chip can be used to hybridize to cellular RNA, and measure the expression rates of a large number of genes in a cell.

TRANSGENICS

Cloning in Animals

Animals have two different classes of cells: germ line and somatic. Only alterations in the germ cells can be transmitted to future generations. However, some forms of somatic cell gene therapy can be useful in treating patients. For example, people with cystic fibrosis can receive the cloned CFTR gene in a nasal spray, which then infects the lining of their lungs and improves lung function. The infected (transformed) epithelial cells eventually are lost, however, during normal processes of tissue growth, so the treatment needs to be repeated.

Germ-line cloning. There are two basic ways to clone a gene in the mammalian germ line.

- Inject a DNA fragment containing your gene into the nucleus of a fertilized egg (germ-line gene cloning) or of body tissues in a mature host (somatic gene cloning). The DNA gets taken up at random somewhere in one of the chromosomes.

An example of germ-line cloning is the injection of angiopoietin cDNA into the fertilized mouse egg.

(An example of somatic cloning is gene therapy for cystic fibrosis: inhale a vector containing the CF gene, which gets incorporated into the DNA of cells lining the lungs. Somatic cloned genes are notinherited by offspring.)

- Put a transgene containing positive and negative selective markers into an embryonic stem cell tissue culture. The ES cells take up the gene by homologous recombination, replacing the host allele. The ES cells now are injected into a blastula of an unrelated host, and some of the next generation progeny arise from the ES cells. To learn more about this, read Capecchi's article on Targeted Transgenes, on reserve.

Once you have cloned a valuable gene into an animal, say a sheep that makes insulin in its milk, how can we produce lots of identical progeny quickly without sexual reproduction? This is what the popular press calls"cloning."

But when you create a clone, how do you know it worked? You need to know:

- *DNA*: Did the transgene really incorporate into the genome? Where?
- *RNA*: Is the RNA expressed? (Or prevented from expression, by a null allele?)
- *Protein*: Is a desired protein expressed?

Cloning in Plants: Plants don't have separate somatic and germ cells, so creating transgenic plants is easier than creating transgenic animals. Techniques include the use of microprojectile bombardment (the"gene gun") and the use of Agrobacterium tumefaciens.

Chapter 5

Biotechnology and Agriculture

REVOLUTIONIZED RESEARCH

Biotechnology has also revolutionized research activities in the area of agriculture which includes the following:

- Plant cell, tissue and organ Culture
- Genetic engineering leading to transformation followed by regeneration of plants to give transgenic plants carrying desirable, traits like disease resistance, insect resistance and herbicide resistance; eventually this may also be used for increasing photosynthetic efficiency, nitrogen fixing ability, improved storage proteins, hybrid crops, crops for food processing etc.
- Somatic hybrids between sexually incompatible species, Permitting transfer of desirable traits from wild or unrelated crop species to ,our crop plants
- Transgenic animals produced in mice, pigs, goats, chicken, cows, etc; it is suggested that some of these will eventually be used as bioreactors to produce drugs through their milk, blood or urine this area has sometimes been described as molecular farming.

The farmers in India are looking at the GM seeds, biofertilizers, biopesticides from which they can expect more return on their investments and also increase productivity. The farmers are now getting a premium for the organic produce. With this now they are able to export their produce at much higher price. Lot of research initiatives are also being undertaken in the field to explore new technologies and new tie ups and joint ventures are being set up and are getting approvals from government for exploiting the technologies for the betterment of farmers and obtaining decent returns roponents of free trade argue that liberalization of agriculture will increase food production and improve the economic conditions of farmers around the world. Real life experience, however, is telling a vastly different story as the liberalization of agriculture unfolds. The entry of transnational corporations in the agricultural sector has created devastating conditions in India, leading many small and marginal farmers, for example, to commit suicide due to economic distress.

The entry of transnational corporations in the sector has moved the focus of agriculture from self sufficiency to making profits. On the one hand, reform policies have ended subsidies provided to Indian farmers, leaving them at the mercy of market forces. On the other hand, these policies have increased incentives for large agribusiness companies to farm. Such is the level of disbandment of farmers in India that, as Devinder Sharma, a well known food policy analyst writes:

"In Andhra Pradesh, the state which leads the country in ushering in progressive policies to improve agriculture, chief minister Chandrababu Naidu has launched the Vision 2020 programme that aims to reduce the number of farmers from the present 70 per cent to 40 per cent. While what happens to the remaining 30 per cent of the farming force is none of his concern, the State continues to record the highest suicide rate among the farming community in the country. Such is the government's apathy that instead of resurrecting the policies to help farmers tide over the crisis of livelihood security, chief minister has been talking of deputing psychiatrists to advise farmers not to commit suicide!"

The impacts of liberalization in India are especially important because over 70% of India's population, directly or indirectly, is engaged in the agriculture sector. Add to this the fact the India is home to nearly a third of the world's poor, and one can begin to realise the numbers of people that trade liberalization, especially of the agricultural sector, impacts.

In their pursuit of profits, transnational corporations are eager to push untested technologies and models regardless of the human, environmental and social damages they may cause. India is rapidly moving towards a cash crop economy, at the expense of self-sufficiency. Large multinationals like Monsanto are also pushing the Gene Revolution (biotechnology), as the successor to the Green Revolution, in spite of largely unanswered questions about the impacts of such a technology. There are grand yet unproved claims that biotechnology is the answer to feeding the hungry in the world. Not only are the unanswered questions problematic, biotechnology in agriculture itself is designed so that corporations gain full control over the agricultural system from local farmers.

Ironically, the hunger problem in India is not one of food production but rather food distribution and storage. If left to the free market, as the WTO would like us to do, there will really be no food at all for the poor since we will export all the cash crops and the imported food will be too expensive for the poor to afford!

What is required is food security and achieving food security will involve consciously planning for the needs of the farmers in India, not abandoning them as has been the case since 1991 when the Indian government began to liberalize the economy. It will also involve guaranteeing genuine land reforms for most Indians, especially in the light that land holding size has been steadily declining over the years.

This issue area looks at the problems surrounding liberalization and corporate agriculture in India, and the attempts to turn back the tides of corporate globalization.

GM SEEDS

The challenge of producing more food grains to feed the ever increasing population of India that has already crossed one billion mark with less resources has bought companies like Mahyco, Monsanto, Syngenta, ProAgro, Advanta to invest in GM crops. It was in 2002 a joint venture between Mahyco and Monsanto called Mahyco- Monsanto Biotech Ltd got the green signal from the Government of India for the commercial production and sale of Bt cotton (Bollgard) in six southern states of India.

A lot of awareness campaigns have to be conducted to reach out to the farmers to brief them about the benefits of using seeds that are resistant to pests, diseases, herbicides, and crops which are tolerant to drought, cold, salinity and other harsh environments. This will bring in confidence among the farmers as well as the industry people.

BIOFERTILIZERS, BIOPESTICIDES

In addition to GM seeds, the farmers are also looking at biofertilizers, biopesticides to get more benefits. Now the farmers are using formulations based on Bt, viruses like NPV, and GV, as well as neem-based pesticides. To meet the increasing demand, the industry has to scale up investments in biofertilizers and biopesticides. Conservative estimate shows that the 10 per cent saving through the use of biofertilizers will result in an annual saving of 1.094 million tons of nitrogenous fertilizers costing around Rs 550 crore.

BIOFUEL

Looking at the opportunity in biofuel sector, the central government has taken an initiative to promote this sector in a big way. The total consumption of ethanol-blended petrol is expected to be 4.6 million tons per year. This sector not only helps sugarcane farmers, as cane is used as raw material for production of ethanol, but also helps in building up the oil security apart from benefiting the environment. It can save foreign exchange to the tune of Rs 80,000 crore, as India imports about 70 per cent of its requirement of crude oil.

USE OF BIOTECHNOLOGY IN AGRICULTURE

The development of Biotechnology has opened many vistas in the field bio-agricultural and bio-medical sciences. The Department of Agriculture and Cooperation has appointed a Task Force on Application of Biotechnology in Agriculture under Dr. M.S. Swaminathan and the

The inter-ministerial consultations on the recommendations of the Task Force have been finalized and report is being examined for final decision. The Department of Agriculture and Cooperation is also placing its emphasis on the following subjects.

CAPACITY BUILDING AND TRAINING OF MANPOWER FOR POST RELEASE MONITORING

The focus of strengthening capability till now has been on bio-safety and environmental safety infrastructure before commercial release of GM crops. With genetically modified cotton already released for commercial cultivation in the country and more crops being in pipeline and the fact that trade in genetically modified food grains will be increasing in future we would be required to deal with GM seeds/crops on a much larger scale. This, inter-alia, calls for upgradation of the post-release monitoring infrastructure which is the responsibility of Department of Agriculture and Cooperation.

In particular, there is a need for capacity building for training of manpower in advance techniques of risk assessment and management of GM crops particularly detection and analysis of LMOs, inspection, monitoring handling of GMO material, quarantine, issues relating to segregation, identity preservation and strengthening of institutions addressing issues of certification. The training of extension personnel and personnel belonging to enforcement agencies like quality enforcement officials of Department of Agriculture of States and personnel from other organizations like Plant Quarantine and Custom officials as well as farmers will go a long way in ensuring post release monitoring and bio-safety.

QUALITY CONTROL OF GM SEEDS AND CROPS

The present Seed Act does not deal with the quality control of GM seeds as they are generally not notified. It is hoped that the new Seeds Act will take care of this issue however, action against spurious GM seeds can always be taken under the Seeds Control Order and the EPA rules. As the GM seeds are very costly and sometimes farmers have been cheated, there is a strong urgency to develop protocol for determining the presence or absence of genes incorporated in the cultivars.

Similarly, the procedure for testing of GM Seeds in the laboratories as well as in the fields is to be developed, strengthened and established with the State Government infrastructure. Strengthening of Seed Testing Laboratories will be undertaken in this context.

CURRENT USES OF BIOTECHNOLOGY IN AGRICULTURE

Genetic engineering can be used to modify the genetic compositions of plants, animals, and microorganisms. The number of genes that have been isolated and are available for transfer is growing daily. Currently, the technology is used primarily to modify crops, although a number of other applications are in the wings.

Like other products, genetically engineered products undergo a period of research and development before they are ready for commercial release. Many products never emerge from the research and development pipeline. While this is true for almost any technology, genetic engineering is turning out to be more difficult and more expensive than early proponents expected.

Although in the early 1980s biotechnology was touted as a miracle technology that was going to usher in a new era of agricultural abundance with minimal harm to the environment, the initial set of products has proved modest. Some of the most important commercial applications of biotechnology are discussed below.

Engineered Crops

The most widespread application of genetic engineering in agriculture by far is in engineered crops. Thousands of such products have been field tested and over a dozen have been approved for commercial use. The traits most commonly introduced into crops are herbicide tolerance, insect tolerance, and virus tolerance.

Herbicide Tolerance

Case Study: Soybeans

Herbicide tolerance allows crops to withstand otherwise lethal doses of herbicides, which are chemicals that kill plants. Some herbicides kill virtually all plants and cannot be used on crops.

By offering crops tolerant to herbicides, chemical companies can expand the market for their products. Indeed, the major developers of herbicide-tolerant plants are companies that sell herbicides. The current set of commercially available herbicide-tolerant crops is tolerant to three herbicides based on three active ingredients: bromoxynil, glyphosate, and glufosinate.

Insect Tolerance

All of the commercially available insect-tolerant plants contain a version of the toxin Bacillus thuringiensis (Bt), which is found in nature in soil bacteria. Bt toxins are highly effective for many pest organisms, like beetles and moth larva, but not toxic to mammals and most other nontarget organisms. A major concern among farmers and environmentalists is that wide use of Bt crops will lead to the rapid development (over the course of perhaps as few as three to five years) of resistance to the toxin. If resistance develops, the Bt toxin will be useless as a pesticide. In this case, the environmental benefits of the product will be short lived.

Loss of Bt efficacy will affect those who currently use the engineered Bt crops, but also many other farmers who use Bt in its natural bacterial form, usually as a spray. These other farmers include those who grow food organically and those who use Bt as part of integrated pest management (IPM) plans. Natural Bt sprays are a valuable mode of pest control for these farmers.

Organic farmers and others who rely on Bt question whether the companies who sell the Bt crops have the right to use up this resource guided only by commercial calculations. UCS considers Bt to be a public good that should be reserved for everyone.

Virus Tolerance

The third major application of biotechnology to crops is virus tolerance. These crops contain a gene taken from a virus. By a process that is not well understood, plants that produce certain viral proteins are able to fend off infections by the viruses from which the proteins were taken. Two virus-tolerant crops are currently approved for commercial use, papaya and squash.

The squash, which is resistant to two viruses, is currently off the market. Although it is difficult to get information on why products are not on the market, it is possible that the squash did not perform well enough in the field to capture market share.

Other Engineered Products

Many other genetically engineered products have been envisioned, but only a few have so far come to market. Several tomatoes engineered to delay ripening have been approved for commercial use. In some cases, delayed ripening just prolongs shelf life. But for the FlavrSavrTM, the objective was to increase the time on the vine without softening, producing a transportable, tasty winter tomato. After a highly publicized rollout, the FlavrSavrTM too is off the market. The problem appears to have been with transportability rather than taste.

On the livestock side, a drug has been produced for dairy cows-recombinant Bovine Growth Hormone (BGH) or Bovine Somatotropin (BST)-by engineering a bacterium to contain the gene for the hormone. The drug is administered to cows to increase milk production, despite the chronic oversupply of milk in the United States. A highly controversial product, BGH is currently used on about a third of the U.S. dairy herd.

An interesting product not related to agriculture is a rabies vaccine intended for use on wild raccoons. In this case, genetic engineering was used to construct a"hybrid"virus made up of a component of the rabies virus inserted into an unrelated"carrier"virus. The resulting virus confers immunity to rabies but poses no danger of causing the disease. Baits laced with the vaccine have been distributed in many parts of the eastern United States in attempts to combat rabies in wild raccoon populations. The vaccine has been approved by the U.S. Department of Agriculture, despite suspicions that it has been only marginally, if at all, effective. Early studies on efficacy failed to demonstrate that the product could control rabies in wild raccoon populations. Data from more recent studies are being withheld from the public as confidential business information.

Agricultural Biotechnology Research Projects

The following lists some of the organisms being engineered by agricultural researchers and indicates which products are commercially available and which are not. For up-to-date information about what is on the market, see our guide toGenetically Engineered Foods Allowed on the Market.

Genetically Engineered Livestock and Poultry

Animals Engineered for Leaner Meat

Status: No livestock engineered for leaner meat is currently near commercialization. Research done early in the 1980s to genetically engineer leaner pigs failed because of unacceptable side effects, including low fertility, arthritis, and impaired immune systems.

Some low level of research activity may still be underway. Pigs have been engineered with roundworm genes to produce their own omega-3 fatty acids, but have not been commercialized.

Animals Engineered as Drug-Production Facilities

Status: Goats and sheep have been engineered to secrete bioactive molecules into their blood, urine, or milk. Companies are in the process of developing commercial enterprises based on these animals. So far, none of the drugs is on the market. It is likely that producers will want to slaughter the animals for food after they are no longer useful for drug production.

Animals Engineered as Sources of Transplant Organs

Status: Commercial entities are engineering pigs so that their organs will not be rejected by human transplant recipients. So far, the organs are not commercially available. It is likely that producers will want to use the carcasses of donor pigs as food.

Animals Engineered for Disease Resistance

Status: Chickens and turkeys have been engineered to resist avian diseases, but not commercialized.

Genetically Engineered Fish and Shellfish

Status: Fish and shellfish have been engineered to cause changes in hormones that accelerate growth in several laboratories, but so far not commercialized in the United States.

Genetically Engineered Plants Eaten Whole as Food

Status: Many plants have been commercialized, including tomatoes and squash and commodity crops like corn and soybeans. Most have been engineered for one of three traits: herbicide tolerance, insect resistance, or virus tolerance.

Genetically Engineered Fibre Plants

Status: Genetically engineered cotton has been approved for commercial use.

Engineered Insects Used in Agricultural Systems

Status: No engineered insects have been approved for commercial use. An engineered predatory mite has been field tested in Florida. Researchers have engineered honeybees and other beneficial insects to tolerate pesticides.

Engineered Microorganisms Used as Pesticides

Status: Several bacteria engineered to enhance their ability to kill or repel pests have been approved for commercial use. These products are used as pesticides in agricultural fields and gardens.

Food Processing Aids Made from Engineered Bacteria

Status: Bacteria have been genetically engineered to produce rennet, an enzyme important in making cheese. Genetically engineered rennet (chymosin) is approved for commercial use and widely used by U.S. cheese processors.

Animal Drugs Made by Engineered Bacteria

Status: Recombinant Bovine Growth Hormone (BGH) made from engineered bacteria is used commercially to stimulate milk production in cows. Currently this hormone is being used in about a third of the nation's dairy herds.

AGRICULTURE BIOTECHNOLOGY FOOD

DISCOVERIES IN BIOTECHNOLOGY

Biogen's Avonex is approved for the treatment of multiple sclerosis. The discovery of a gene associated with Parkinson's disease provides an important new avenue of research into the cause and potential treatment of the debilitating neurological ailment.

The Age of biotechnology arrives with"somatostatin"- a human growth hormone-releasing inhibitory factor, the first human protein manufactured in bacteria by Genentech, Inc. A synthetic, recombinant gene was used to clone a protein for the first time.

1978: Genentech, Inc. and The City of Hope National Medical Centre announce the successful laboratory production of human insulin using recombinant DNA technology. Hutchinson and Edgell show it is possible to introduce specific mutations at specific sites in a DNA molecule.

1979: Sir Walter Bodmer suggests a way of using DNA technology to find gene markers to show up specific genetic diseases and their carriers. John Baxter reports cloning the gene for human growth hormone.

1980: The prokaryote model, E. coli, is used to produce insulin and other medicine, in human form. Researchers successfully introduce a human gene - one that codes for the protein interferon- into a bacterium. The U.S. patent for gene cloning is awarded to Cohen and Boyer.

1981: Scientists at Ohio University produce the first transgenic animals by transferring genes from other animals into mice. The first gene-synthesizing machines are developed. Chinese scientists successfully clone a golden carp fish.

1982: Genentech, Inc. receives approval from the Food and Drug Administration to market genetically engineered human insulin. Applied Biosystems, Inc. introduces the first commercial gas phase protein sequencer.

1983: The polymerase chain reaction is invented by Kary B Mullis. The first artificial chromosome is synthesized, and the first genetic markers for specific inherited diseases are found.

1984: Chiron Corp. announces the first cloning and sequencing of the entire human immuno-deficiency virus (HIV) genome. Alec Jeffreys introduces technique for DNA fingerprinting to identify individuals. The first genetically engineered vaccine is developed. Cetus Corporation's develops GeneAmp polymerase chain reaction (PCR) technology, which could generate billions of copies of a targeted gene sequence in only hours. Scientists find a gene marker for cystic fibrosis on chromosome number 7.

1986: The first genetically engineered human vaccine - Chiron's Recombivax HB - is approved for the prevention of hepatitis B. A regiment of scientists and technicians at Caltech and Applied Biosystems, Inc. invented the automated DNA fluorescence sequencer.

1987: The first outdoor tests on a genetically engineered bacterium are allowed. It inhibits frost formation on plants. Genentech's tissue plasminogen activator (tPA), sold as Activase, is approved as a treatment for heart attacks.

1988: Harvard molecular geneticists Philip Leder and Timothy Stewart awarded the first patent for a genetically altered animal, a mouse that is highly susceptible to breast cancer

1989: UC Davis scientists develop a recombinant vaccine against the deadly rinderpest virus. The human genome project is set up, a collaboration between scientists from countries around the world to work out the whole of the human genetic code.

1990: The first gene therapy takes place, on a four-year-old girl with an immune-system disorder called ADA deficiency. The human genome project is formally launched.

1991: Mary-Claire King, of the University of California, Berkeley, finds evidence that a gene on chromosome 17 causes the inherited form of breast cancer and also increases the risk of ovarian cancer. Tracey the first transgenic sheep is born.

1992: The first liver xenotransplant from one type of animal to another is carried out successfully. Chiron's Proleukin is approved for the treatment of renal cell cancer.

1993: The FDA declares that genetically engineered foods are"not inherently dangerous"and do not require special regulation. Chiron's Betaseron is approved as the first treatment for multiple sclerosis in 20 years.

1994: The first genetically engineered food product, the

Flavr Savr tomato, gained FDA approval. The first breast cancer gene is discovered. Genentech's Nutropin is approved for the treatment of growth hormone deficiency.

1995: Researchers at Duke University Medical Centre transplanted hearts from genetically altered pigs into baboons, proving that cross-species operations are possible. The bacterium Haemophilus influenzae is the first living organism in the world to have its entire genome sequenced.

1996: Biogen's Avonex is approved for the treatment of multiple sclerosis. The discovery of a gene associated with Parkinson's disease provides an important new avenue of research into the cause and potential treatment of the debilitating neurological ailment.

1997: Researchers at Scotland's Roslin Institute report that they have cloned a sheep--named Dolly--from the cell of an adult ewe. The FDA approves Rituxan, the first antibody-based therapy for cancer.

1998: The first complete animal genome the C.elegans worm is sequenced. James Thomson at Wisconsin and John Gearhart in Baltimore each develop a technique for culturing embryonic stem cells.

1999: A new medical diagnostic test will for the first time allow quick identification of BSE/CJD a rare but devastating form of neurologic disease transmitted from cattle to humans.

2000: "Golden Rice,"modified to make vitamin A. Cloned pigs are born for the first time in work done by Alan Coleman and his team at PPL, the Edinburgh-based company responsible for Dolly the sheep.

2001: The sequence of the human genome is published in Science and Nature, making it possible for researchers all over the world to begin developing genetically based treatments for disease.

2002: Researchers sequence the DNA of rice, and is the first crop to have its genome decoded.

2003: The sequencing of the human genome is completed.

WORLD AGRICULTURE

It is imperative to make agriculture more sustainable if we want to preserve this planet and eliminate the possibility of the global famine in the future. Creating a sustainable agriculture system is actually much more difficult than developing the concept of it in theory.

Sustainable agriculture is an issue that many environmental scientists started pondering in recent years. The reason it is so vital is because food is something that people need for physical survival on this planet and agriculture directly deals with how to produce it. Nevertheless, there is a misconception that sustainable agriculture deals only with how to feed the expanding population today and how to produce enough food to meet the dietary demands at present. The difference here lies in the core objective of this branch of environmental science. It is a much broader concept which is concerned with meeting the demand for food in the future. It is a fact that some geographic regions that used to have fertile soil have become deserts due to inefficient agricultural practices that are common in the world today.

Therefore, the fact of food overproduction at present will not seem so exciting if the future generations starve to death not having enough supplies to feed the ever-growing population. Therefore, sustainable agriculture is concerned with developing an efficient environment-friendly food production system that would eliminate a waste of limited natural resources and prevent land from losing fertility while producing adequate yields year after year.

It is still important to produce enough food today but it is imperative that substantial agriculture factor in deterioration in fertility and depletion of natural resources, particularly soil and water. So far we have 3 areas that substantial agriculture is concerned with: meeting the demand for food today, ensuring that the future generations will be able to produce enough food given the ever-present deterioration of soil and water, and taking control of soil and water condition.

It is imperative to make agriculture more sustainable if we want to preserve this planet and eliminate the possibility of the global famine in the future. Creating a sustainable agriculture system is actually much more difficult than developing the concept of it in theory. It presents a real dilemma simply because numerous intertwining macro and microeconomic factors influence the level of quality and form of a product produced by the system. Therefore we have to factor in the complex economic and political environment while theorizing about sustainable agriculture systems. As a result, farmers are forced to produce products that people are willing to buy and that are safe to eat, compete with other producers, and act within political and legal boundaries. By the same token, sustainable agriculture creates a framework that farmers must act within, in other words a set of rules to comply with. This simply emphasizes that agriculture is a difficult business to stay in and to make it more sustainable is to complicate it even further. Nevertheless the level of difficulty involved must not deter us from this task because the potential results will definitely outweigh the problems that farmers might be facing today.

Every year the total area of agricultural land gradually decreases because land is being eaten up by rapid development. At this point in time the US can produce more food than it needs and this can probably go on for several decades but if no radical measures are taken the aggregate supply of food in the global economy will dwindle. One of the potential solutions to this problem would be government support in the form of lower taxes for farmers.

This would help people stay in business longer and concentrate more on environmental problems. In order to make the agriculture more sustainable farmers should utilize natural resources more efficiently. Some advanced cropping techniques must be employed to maintain soil fertility at a certain acceptable level and prevent it from deterioration. They also need to control the amount of fertilizers used in the process thus producing safe to eat and healthy products. Currently, the situation is far from perfect but it is headed in the right direction with the government starting to acknowledge the importance of this issue and providing support to farmers.

Tim Johnson Junior is a freelance writer, CRWA certified resume writer and career coach. Has written over 2000 articles and essays on the subject of Social Issues. Has worked for Essaymart's custom writing department from 2003 to 2005. Currently, Tim is busy helping professionals and executives optimize their careers at a certified Resume Writing firm, ResumeAid.

BIOTECHNOLOGY IN INDIA

Cloning is also not a dirty word in India."Hinduism will not have any major conflicts with engineered life forms of any kind because the tradition has always had multiple life forms and considers any and all of them as co-travellers on the Möbius strip,"said Venkatraman.

LONAVLA, India

In India, Hindu culture trumps all. And although India is a growing hub of technological and biological influence, Hinduism dominates even the sciences. India is ranked 37 among the 82 countries assessed by the World Economic Forum's Global Competitiveness Report for the"state of their information technology system and its effects on economic growth and productivity."

Roughly 300,000 engineers graduate from Indian colleges and universities each year. Multinational companies are taking advantage of the talent pool by making major high-tech investments, such as Microsoft's plan to spend $1.7 billion and hire 3,000 employees in India over the next three to four years.

India's biotech industry is also on the rise, with 500,000 doctors and nurses entering the workforce annually. Stem cell research in both the public and private sectors has grown considerably over the past few years in India, where politics or faith has not hindered its expansion. As a result, India is home to not one but three national stem cell research facilities.

In Western nations like the United States, however, stem cell research is a hot-button issue. Just a public discussion of the research has triggered furious protests and stirred up government officials. Not so in India, where the Hindu-influenced worldview pervades scientific progress and everyday discourse.

Hinduism, for its part,"doesn't share the moral skittishness sometimes displayed by Western Christian thought,"said Arvind Sharma, the Birks Professor of Comparative Religions at Montreal's McGill University. If no life is destroyed when taking stem cells from an aborted fetus, and the purpose is not evil, it would not disturb their morality, he said.

To keep things on an even keel, secular committees issue national directives. In 2004, the Central Ethics Committee on Human Research of the Indian Council on Medical Research circulated ethical guidelines on how to conduct stem cell research.The Draft Guidelines on Stem Cell Research/ Regulation stresses that"termination of pregnancy for obtaining fetus for stem cells, research or for transplantation is not to be permitted."Additionally,"no embryo can be created for the sole purpose of obtaining stem cells."In 2000, a report on"Ethical Guidelines for Biomedical Research on Human Subjects,"which dealt with genetic screening, was released.

Recommendations such as these are totally in character for the general milieu, said Sharma."Most moral issues don't come into the public discourse but remain private."Using the example of another bioethical controversy that is contentious in the West, he added,"People deal with issues like euthanasia in the context of their families."

India is officially a secular republic, home to the largest number of Hindus and Muslims in the world."Nearly every Indian, regardless of religion, is Hindu-thinking and lives according to Hindu culture and philosophy,"said Ram Surat, a Christian convert getting his divinity degree at the Union Biblical Seminary, Pune. For Hindus, this philosophy translates to a respect for all life, a belief in an immutable soul and the body as a vessel. Even Christians - a growing population in India - do not have as strong criticisms of biotechnology as their Western counterparts. The reason is that Hinduism casts a long shadow even over other religions.

Few Christians in India talk about such issues, said Selva Raj, the Stanley S. Kresge Professor of Religious Studies at Albion University in Michigan."Indian Christians are much more interested in how to live and coexist with people of other religions."

"Life and death are not points in a line. It is a Möbius strip,"said Shridhar Venkatraman, an engineer in Chennai who lived for 10 years in the United States."All living things work towards escaping this cycle,"and so life and death are personal issues. The news describes discoveries in science as well as the furor they cause in the West. But in general, the discussion is digested silently."Bioethics is only discussed by the very few elite,"according to Dhruv Raina, a professor at the Zakir Husain Centre for Educational Studies at Jawaharlal Nehru University in New Delhi."In general, the well-being of people prevails over ideas of danger,"said Raina, who researches the relationship between science, societies, values and culture.

Hinduism, itself, is not a monolithic entity."Unlike Christianity, Hinduism is not a codified religion with a single papal authority to pontificate on every subject,"said Jayanthi Iyengar, a practitioner of the Art of Living, in Pune."You won't find a position on these issues like the one the Catholic Church has on abortion or genetic modification,"Iyengar said.

Hinduism has a fulcrum of pragmatism,"according to Lalitha Khanna, a researcher with a Delhi-based think tank."What is good for making a better world is condoned, even eagerly embraced. Stem cell research, therefore, doesn't bring out the fierce opposition that Christians in the West probably experience and evince,"she added. Religious mandates would be out of place here."Every sect and subsect has a guru of its own and will not follow the religious directions of another,"said Khanna.

Cloning is also not a dirty word in India."Hinduism will not have any major conflicts with engineered life forms of any kind because the tradition has always had multiple life forms and considers any and all of them as co-travellers on the Möbius strip,"said Venkatraman.

"We are culturally desensitized to the possibility of the existence of such things,"added Sharma. Case in point: The Hindu god of good beginnings, Ganesha, is human with an elephant's head; the god Vishnu came to earth as a narasimha - half man, half lion.

Most Indian children learn these stories growing up, regardless of religion."At the level of practice, I think Indian Christians are pretty pragmatic in their use of technologies,"said Rowena Robinson, an associate professor of sociology at the Indian Institute of Technology, Bombay."I am not sure if the ideological implications cause much wringing of hands,"she said.

It is wrong to think science and religion are in conflict in India, added Victor Ferrao, a doctoral student at the Jnana-Deepa Vidyapeeth seminary in Pune. In his hometown of Goa, Ferrao leads a community science-and-religion dialogue group."Developments in science make the dialogue urgent,"he said,"but science and religion are correlational."

INPUT FOR AGRICULTURE

Improves fruit quality and prevents some plant diseases: The quality of fruit depends on the amount of calcium. Calcium is nearly always only marginally sufficient and often deficient in developing fruits. Calcium moves very slowly, if at all, from one plant part to another and fruits at the end of the transport system get too little. Calcium must be constantly available to the roots.

India ranks second on the basis of population in the world. Agricultural land utilised by the burgeoning population, the cultivable land resource is shrinking day by day. To meet the food, fibre, fuel, fodder, and other needs of the growing population, the productivity of agricultural land has to be increased rationally. This requires the use of all resources judiciously. In India, the mineral gypsum is mostly used in the manufacture of cement, fertilizer, plaster of Paris, ceramics and distemper. Smaller quantity is used as soil conditioner, for carving and statuary purposes. India has huge resources of natural gypsum of the order of 1120 million tonnes, of which recoverable reserves are estimated at 237 million tonnes. Over 95 per cent of the natural gypsum come from Rajasthan.

Gypsum is chemically calcium sulfate dihydrate ($CaSO_4.2H_2O$). When dissolved in water, it yields calcium ions (Ca_2+) and sulfate-sulfur ions (SO42-). Both of these ions are essential major nutrients for growing plants. In addition to this, calcium also plays a vital role in establishing and maintaining good chemical balance in soil, water and plants. Gypsum is one of those rare materials that perform in all three categories of soil treatment: an amendment, conditioner, and fertilizer.

The usage of gypsum in agriculture can be grouped into following heads:

- *Reclaims soil sodicity*: The sodic soils have exchangeable sodium percentage (ESP) more than 15, it must be lowered for soil improvement as well as better crop growth. The calcium supplied by gypsum replaces the sodium held at the clay-exchange sites. The replaced sodium can be leached from the soil as sodium sulfate to an appropriate sink.
- *Decreases the toxic effect of NaCl salinity*: Calcium from gypsum has a physiological role in inhibiting the uptake of sodium by plants. Thus plant are mitigate the sodium toxicity, which is more pronounced in the salt affected soils.
- *Decreases pH of sodic soils*: The sodic soils are characterised by the high soil pH (>8.5). Gypsum lowers the high pH of sodic soils or near sodic soils to 7.5 to 7.8. These values are in the range of acceptability for growth of most crop plants. Increased calcium uptake by roots when gypsum is applied can decrease the pH of the rhizosphere.
- *Creates favourable soil EC*: The more EC value of the soil is undesired for the crop growth. The high EC of soil are due to fertilizer application as well as by the weathering of soil minerals. Gypsum, being

readily soluble, results in proper buffered solute concentration (EC) in soil to maintain soil in a flocculated state.

- *Makes excess Magnesium non-toxic*: Calcium is essential to the biochemical mechanisms by which most plants nutrients are absorbed by roots. Without adequate calcium, uptake mechanisms would fail. In soils having unfavourable calcium magnesium ratios, such as serpentine soils, gypsum can create a more favourable ratio. Thus render Magnesium non-toxic.
- *Decreases heavy-metal toxicity*: Calcium plays a role of regulator for balance of micro-nutrients, such as iron, zinc, manganese and copper, in plants. Calcium prevents excess uptake of many of them; and once they are in the plant, calcium keeps them from having adverse effects when their levels get high. Calcium in liberal quantities helps to maintain a healthy balance of nutrients and non-nutrients within plants. Gypsum also removes excess of Boron from sodic soil. It also regulates non-essential trace elements.
- *Improves soil structure*: Gypsum provides calcium, which, flocculate clays in acid and alkaline soil. A flocculated clays form friable soil with improved soil structure and tilth. It also allows for deeper, healthier root development and water penetration.
- *Prevents crusting of soil*: Gypsum application can decrease and prevent the crust formation on soil surfaces which result from rain drops or from sprinkler irrigation on unstable soil. It can prevent crusting that result when acid soils are limed and the gypsum is coapplied with the lime. Improves compacted soil: Gypsum can break up compacted soil and decrease penetrometre resistance. Combination with organic amendments also helps, especially in preventing return of the compaction. The soils applied with gypsum have low bulk density.
- *Makes slightly wet soils easier to till*: Soils that have

been treated with gypsum have a wider range of soil moisture levels where it is safe to till without danger of compaction or deflocculation. Even a liberal application of gypsum is a good procedure for starting a piece of land into no-till soil management or pasture. Gypsum also improves the ability of soil to drain and not become waterlogged due to a combination of high sodium, swelling clay, and excess water.

- *Stops water run-off and erosion*: Gypsum improves water infiltration rates into soils and also the hydraulic conductivity of the soil. The use of gypsum can decrease wind and water erosion of soil.
- *Improves swelling clays*: Gypsum can decrease the swelling and cracking associated with high levels of exchangeable sodium on the montmorillonite-type clays. As sodium is replaced by calcium on these clays, they swell less and therefore do not easily clog the pore spaces through which air, water and roots move.
- *Increases water retention in soil*: Gypsum when applied to sodic soil reduced the levels of exchangeable sodium resulted in increase in water retention. The improved soil structure help in the more retention of water. Increases the stability of soil organic matter: Gypsum is a source of calcium responsible for the binding of soil organic matter to clay and gives stability to soil aggregates.
- *Improves low-solute irrigation water*: Gypsum is used to increase the solute concentration of low-solute water used for irrigation. Irrigation water from rivers that no longer have sources of leachable salts either penetrates poorly into soil or causes soil particles to degrade which results in low-water penetration. The problem can be corrected with surface-applied gypsum or application to the irrigation water.
- *Improves water-use efficiency*: Gypsum application increases water-use efficiency of crops. Improved

water infiltration rates, improved hydraulic conductivity of soil, better water storage in the soil all lead to deeper rooting and better water-use efficiency.

- *Efficiently use low quality irrigation water*: Use of reclaimed municipal wastewater is important for conservation of natural resources. Reclaimed water can be satisfactorily used if amended with gypsum and water-soluble polymers.

Improves fruit quality and prevents some plant diseases: The quality of fruit depends on the amount of calcium. Calcium is nearly always only marginally sufficient and often deficient in developing fruits. Calcium moves very slowly, if at all, from one plant part to another and fruits at the end of the transport system get too little. Calcium must be constantly available to the roots. In very high pH soils, calcium is not available enough; therefore, gypsum helps. Gypsum is used for peanuts, which develop below ground, to keep them disease free. Gypsum helps prevent blossom-end rot of watermelon and tomatoes and bitter pit in apples. Gypsum is preferred over lime for potatoes grown in acid soils so that scab may be controlled.

- *Decreases loss of fertilizer nitrogen to the air*: Calcium from gypsum can help decrease volatilization loss of ammonium nitrogen from applications of ammonia, ammonium nitrate, urea, ammonium sulfate, or any of the ammonium phosphates.
- *Keeps clay off tuber and root crops*: Gypsum can help keep clay particles from adhering to roots, bulbs and tubers of crops like potato, carrots, garlic and beets.
- *Helps Earthworms to Flourish*: A continuous supply of calcium with organics is necessary for the habitat of earthworms that improve soil aeration, improve soil aggregation and churn the organic matter with inorganic fraction of soil.

INDIA FARMERS

As each gardner is growing twice the table food required

by his family or more, let the surplus be split. Half for fresh seed and half for the local market. The independence and security the regular production of surplus food engenders cannot be valued in currency.

The promise is the same everywhere. Buy our patented seeds at twice the price of others, buy our chemical fertilizers and pesticides and you will prosper like the government subsidized American farmers. These corporate parasites know how to sell in any language. The cost of the promised prosperity which seldom occurs without government subsidy and requires annual borrowing, as with American farming, is more than a poor farmer can afford. But corporate greed convinces him to gamble his future and his life on false corporate hope.

Working hand in glove with the corporate scum are the local money lenders, sometimes taking 5% interest per month and sometimes the entire harvest if drought or floods do not beat Mr. Moneybags to the harvest. India could not have chosen a worse model to modernize its agriculture than American corporate farming, which has already destroyed more value than it has produced, excluding corporate profits of course, because corporate balance sheets are all that matter in this world. Raped lands and environments be damned.

People and the people in government especially, easily lose sight of the fundamental importance of agriculture to the total national economy. A nation which does not feed its people is forever dependent upon the good will of people who cannot be trusted. Even the rich nations have boarded this leaky boat and they will pay dearly for so doing and are already paying dearly, in terms of food quality, environmental damage and disease.

If a nation is to have healthy people, requiring a minimum of medical attention, it must have health producing food, not the plastic the corporations peddle in the name of high yields and damage resistance. To have health building food, grow in rich organic soil 100% chemical free. In the story I read, a 31 year old farmer borrowed $9 for a bottle of pesticide which he drank to kill himelf. So the manufacturer made five dollars to create a widow and two small fatherless children. Congratulations on your keen business acumen, corporate parasite. Win - lose marketing at its finest.

Indian government must discourage all farming and switch to organic gardening to continuously enrich the soil upon which a combination of subsistence and high value cash crops are grown. Competing with government subsidized, environmentally destructive, disease producing corporate agriculture is like running a marathon on your hands and knees. You might finish the race but you will win no prize. We are talking about a sustainable foundaion for a national economy. It deserves more thought and planning than where to build the next chemical disaster.

To have rich and healthy soil as a foundation for prosperity it is essential to compost mineral rich plants and manures, green and vegetariabn animal wastes. The best source of mineral rich plants are seacoasts. The next best source is former cotton fields planted in alfalfla or a similarly deep root crop which mines the deep minerals in sub soil. A government which will subsidize small scale agriculture will make strict conditions. A gardener must produce twice the food his family requires, so a partial failure allows his family to eat. Crops sprayed with a mixture of kelp or seaweed and fish emulshion are drought resistant, insect resistant (few insects eat fish or kelp), disease resistant and yield more plant and fruit than anything corporate agriculture will ever have to offer.

Build raised gardens close to the house or shanty where it is natural to keep a close eye on them. Make them any length and four feet wide, waist high. Fill the bottom half with sand or stone, the top half or one third with topsoil and compost. Build with brick, block or stone which will not wash away with heavy rain. Use either a square foot method or third metre, in the case of a one metre wide bed. The idea is an easy reach from each side to the centre of the bed. By growing food and high value cash crops in these raised beds, one has purchased one time flood insurance. Small pipes at the bottom of these beds allow the bed to drain by gravity. The soaked soil will drain. If the rain is severe or prolonged, spread inexpensive covers over a sturdy frame above the plants in the bed. In Winter or cold weather, clear plastic covers will heat the bed and prevent crop freeze overnight. The thermal mass of such elevated beds acts as a thermal flywheel, maintaing desired temperatures.

As each gardner is growing twice the table food required by his family or more, let the surplus be split. Half for fresh seed and half for the local market. The independence and security the regular production of surplus food engenders cannot be valued in currency. This is private, personal, life and health insurance. Cash crops should be both common and coveted medicinal herbs and spices. There will always be a free market for such. Wisely chosen herbs and spices may be dried and shipped around the globe at very high profits. Trends in rich countries will always tell you how to harvest the most cash from a square foot or third metre of crop.

Square Foot Gardening is now so highly developed, one can grow five times the crop volume with the same resources wasted in conventional farming and gardening. Any nation not using this method is wasting resources and labour. Harvest humanithy for record profits and growth, not chemicals and scarce resources. Outlaw the corporate confidence men and the local money lenders. Teach your rural populations sustainable agriculture and become the dominant economic power of the New Age. The wisdom to do so is among your people. When you end farming, you end farmer suicides.

HEALTH FARMS

Health farms have doctors for medical advice and counselors for alternative treatment therapies. They can help you with a treasure house of information on better living. Make the most of your stay by picking up a few tips.

Tired of living in the steel and glass urban jungle? Or feeling a little breathless trying to keep up with those impossible deadlines at work? Maybe your social and personal life is slowly turning into a mess and you know you could do with some more time to yourself.

Modern life comes with its own stresses and strains. We seem to work doubly hard to keep our place and with long working hours, erratic eating habits it's easy to see how you can get burnt out. When you stop to think about it, are you as far way from this as you really think? It seems easy to get absorbed in career life but the downside is when health, relationships or both take a back seat to work.

However there is some hope. Health farms are now filling a felt need for a retreat from the routine. These venues vary in size and location and some are even country mansions with extensive grounds. They provide a cozy ambience and fantastic facilities. Whether you are a frazzled exec seeking a break from the boss or a working mum trying to reconnect to things past, a visit to a health farm can rejuvenate the mind and body.

Here are some key reasons why more and more people are choosing health farms as a short break: A short stay in a health farm is a great tonic providing both physical and mental relaxation. It allows you to dabble in things that you would normally brush aside in your hectic work schedule and it gives moments of solitude and introspection.

A Spa Break can also help you relax and cements closeness, which can help bring the love back into relationships.

Often the most attractive feature of health farms is the spa. The spas offer excellent facilities for recuperation of the tired body and mind. Body massage with aromatic oils, refreshing baths in perfumed water and mud treatments can help improve blood circulation and ease muscle stiffness.

Skin treatments include exfoliation, scrub with herbal products and manicure. Meditation, kinesiology and yoga classes can help you learn how to cope with stress. Personal consultations are also available on several lifestyle related disorders and the counselors can explain healthier options in food and living.

Taking long walks in the countryside, playing golf; relaxing outings at the pool and tasting delectable cuisine make the whole experience a memorable one.

A typical Spa 2 day package could include: A body massage, facial, a complete programme of exercise and relaxation class, a balanced meal programme, luncheon on day of arrival and departure, breakfast and dinner as well as unlimited use of the Spa facilities, hydro pool, luxurious sauna and steam rooms. There are many health farms in the UK and Ireland. Ragdale Hall, Champneys and Galway are some of the popular health farms. The Utopia Spa located in Kent is also proving to be increasingly popular too.

Blessed with verdant greenery all around, these farms attract the rich and famous with many valued added services. With the growing popularity of the Internet, several services for planning a holiday at a health farm. And lets not forget, Health farms are not just for the health conscious. They make a great gift idea for a loved one too.

Here are a few simple things that you should keep in mind while planning a holiday at a health farm: Check availability of accommodation and always pre-book your stay. Forward instructions about any specific preferences to the Spa - this applies especially to any food and treatment choices. Check the residential rules at the venue, but most of all relax and enjoy yourself. Health farms have doctors for medical advice and counselors for alternative treatment therapies. They can help you with a treasure house of information on better living. Make the most of your stay by picking up a few tips.

Whether you choose a day spa break or stay for a few days, health farms are a great way to beat the circus of the daily grind. It is an invitation for personal indulgence to rejuvenate and de-stress. If you want to do your body and mind a favour as well as spend some quality time alone with a friend or partner then a Spa break could be just what the doctor ordered.

LOCAL FARMING

Two things children love is to play in the dirt and to be active. Letting them plant, care for and harvest their own little garden accomplishes both. The underlying benefit to parents and food retailers is that they are learning.

This week, I spent a couple of hours working at my CSA

farm. I harvested two rows of fingerling potatoes before the thunderstorms set in and I was shooed out of the field by the farmer, Mary LaFrance. I am a member of Lakeplain Prairie Organic Farm; the only certified organic CSA in Wayne County, Michigan. For those of you who aren't familiar with CSA farms, they offer a subscription based service for fresh, good tasting, local and organic food. Subscribers to this service pay a seasonal fee and agree to work a specified number of hours over the course of the growing season.

As a CSA member you develop a relationship with a local farmer, receive a weekly share of fresh-picked great tasting local food, and cultivate friendships in your local community. You also develop a strong sense of satisfaction that you know where your food was grown, when it was harvested and the distance it traveled to your plate.

At first glance, you may think food retailers are in direct conflict with CSA Farms. Admittedly, if I am getting all my fruits, vegetables and flowers from my CSA membership I am no longer buying them from my local grocer. If that's the case, why would my local grocer want to promote community supported agriculture? If you subscribe to the theory of economic abundance, there are enough resources for everyone. With a bit of cooperative support and some creative marketing, food retailers and CSA Farms can foster economic abundance in your local community.

Cross Marketing with your local CSA

CSA Farms often supplement their earnings by reserving a portion of the farm for direct sales to local retailers. By taking advantage of this freshly picked locally grown organic produce, these stores are not only providing a better product for their customers, but supporting their local farmers.

Don't keep this locally grown produce a secret:

- Label your locally grown produce with the farmers name, location and photo, if available.
- Invite the local farmers into your store for a"Meet the Farmers"day.
- Give out recipes promoting seasonal produce and

provide information about the farm where it was grown and harvested.

Letting your customers know you support local farmers generates community goodwill and keeps dollars flowing among friends and neighbours. Produce not native to your growing region will also still be in demand and should be supplied by your retail store. And remember, your local farmers need to shop, too! Support them and they will in turn support your business.

Co-Sponsor Educational Events

Many CSA Farms offer classes in canning, freezing and preserving fruits and vegetables. Co-sponsoring a CSA educational event is a great way to show your support for the community while reminding residents that you can provide all of their shopping needs not available through their CSA. The strawberry preserves members are learning to make still require sugar, fruit pectin and canning jars! Promoting the event with signage at your retail store further shows your support for the community and your local food network.

Children: Your Next Generation of Local Food Consumers

My CSA Farm, Lakeplain Prairie, has a special garden area just for children. Not only do CSA member parents have an opportunity to complete their service requirement uninterrupted, children have an opportunity to learn about gardening. Two things children love is to play in the dirt and to be active. Letting them plant, care for and harvest their own little garden accomplishes both. The underlying benefit to parents and food retailers is that they are learning. Children will experience a sense of pride and enjoyment when growing and cultivating their own fruits, vegetables and flowers. They also tend to be more willing to taste produce they grew on their own, leading to healthy eating and less risk of obesity. And remember, today's little farmer is tomorrow's busy consumer.

As a food retailer, supporting your local CSA farms makes good business sense. Cross marketing, educational programmes and engaging children are all ways to increase retail sales. Long term benefits to your retail business will come through fostering community goodwill, consumer dietary health and a strong local economy.

ORGANIC FARM

Each incoming request is relayed round-robin style to the next server in the web farm. The View State remains persistent across each different server, yet no special programming or state handling mechanism was employed. In fact, the application developers themselves do not even know the secret.

Act One: The Pledge

Dear Reader, consider yourself warned! The following paragraphs are not for the faint-hearted. Contained herein lies a feat of daring so bold, so extraordinary, so shocking that it mocks your very understanding of the universe. Best you turn away now, lest the fabric of your reality unravel as so many loose threads; leaving you standing bare and raving mad, forsaken in a wasteland of shattered beliefs.

I'm going to take a run-of-the-mill session-based application and turn it into a load balanced web farm before your very eyes - all without making a single coding change to the application. That's right ladies and gentlemen, what you're about to witness is a load balancing technique so ingenious, so perfect that the server - nay the very application itself will not even know it is part of a web farm.

Each and every server in the farm will think it alone is communicating with the client. Each and every server will record a hit not from the load balancer, but from the client itself. And finally each and every server will respond directly back to the client, utterly bypassing the very load balancer that brokered the request to begin with. No special software will be installed on the servers. No ISAPI filters. No plug-ins or other such nonsense will be required.

Understand! This is hardware load balancing I speak of, pure and simple. Hardware load balancing accomplished without the likes of Zeus, Coyote Point or even the F5 BigIP.

This is organic farming; hardware load balancing achieved with nothing more than two ordinary servers. And perhaps, a little bit of magic.

Act Two: The Turn

I have before me two Dell PowerEdge 850s, each with but a single Celeron processor and a mere 512 MB of memory. One could use nearly any kind of hardware. We, of course, have our corporate sponsorships to think of.

With a wave of my magic wand, Debian Sarge is installed on both servers. Do not be deceived by its meek and unassuming nature. Renowed for its robustness to hyperbole and found in all 12 world climate zones, Debian Linux stands ably by to perform any task at my bidding.

After a sprinkle of additions to the source.list file, I shall now utter the magic words to turn these servers into a high availability pair of load directors: apt-get install ultramonkey. Now with another wave of my wand, I apply the necessary configuration elements to enable the fabled load balancing method known as Direct Routing. Observe! The default gateway of each server remains unchanged (and it is not the IP address of the load balancer). Our servers do not even know the load balancer exists. Requests are forwarded with their source IP address intact. Each server replies directly, bypassing the load balancer entirely.

Act Three: The Prestige

Still not impressed? Take heed, for I have but two words that will surely sway even the most skeptical of skeptics. Listen carefully, for I shall not repeat myself. The words are View State. That's right, I have accomplished the implausible, the unthinkable. You'll find no Apaches in this web farm. I present to you Windows Server 2003 and a Microsoft.NET application.

How many in our audience have tried load balancing a.NET application? If so, then you surely have encountered the dreaded"View State is invalid"error. If you"fixed"this error by enabling some type of sticky sessions on your load balancer, I applaud you with a tip of my cap and a knowing wink. An honourable mention of your efforts is in order perhaps, but let's be honest. That's not really load balancing is it? In the end each client is bound to a single server. Stickyness is only excusable should your application require SSL.

Behold! I present to you the World's Greatest Unsticky Organic Farm. Each incoming request is relayed round-robin style to the next server in the farm. The View State remains persistent across each different server, yet no special programming or state handling mechanism was employed. In fact, the application developers themselves do not even know the secret. And should one server be unavailable for whatever reason, our load balancers will automatically detect it and only send requests to the remaining live servers.

Act Four: The Secret

Everyone knows that a good magician guards the secret to his magic.

FACTORY FARMING

For"DES girls", as these offspring were soon labeled, cervical cancer and higher incidence of irregular menstruation, structural malformation of the vagina and cervix, future miscarriages, premature births and infant deaths during their reproductive future.

Everyone has heard the expression"you are what you eat". However, in our drug-and-technology industrial food economy based on factory farming"protein"such as chicken, lamb and beef or even fish, you may discover that you're slightly more than simply what you eat these days. Other"stuff"may be happening.

When a guy's voice begins to migrate towards falsetto and wee breasts begin to form, you want to be asking"what's going on here"? When couples struggle to get pregnant...when a field mouse makes more morphologically sound sperm cells than an adult human male...when 2nd and 3rd generation women develop higher rates of miscarriage and cervical cancer...when infant boys show atrophied and un-descended testicles and undersized penis, then it's time to look to what's inside our food chain such as the slough of man-made estrogens like DES or diethylstilbestrol and growth stimulants used to trick Mother Nature.

Where's The Meat? Getting consistent quality cheap meat seems a basic right to Americans. Once grass land feeding proved inferior to mass-produced feedlot cattle, delivering beef to the American dinner table would never be the same. Bringing tens of thousands of cattle into a confined space, limiting movement, keeping them indoors standing up to their knees in muck and waste meant introducing heaps of antibiotics in order to control disease and keep the cattle alive until slaughtering time. It also meant working out chemical"tricks"to stimulate growth, such as using heavy dosages of female hormone compounds delivered to one and all cattle!

Meat Hormone History - Feeding America Scientifically. Initially, DES or diethylstilbestrol was synthesized in the late 1930s. A University of California poultry researcher discovered after a test that DES produced a curious, if not outright profound effect in male chickens. Treated males were instantly chemically castrated, changing into capons, exhibiting pronounced and juicy breast meat after their diethylstilbestrol injections! Plus, animals could be brought to the slaughterhouse and market significantly quicker...which meant lowered costs and increasing profits! The dinner bell rang loud and clear. Treat chickens with DES and you've got fast-producing, lower cost chicken on America's dining table.

All was hunky-dory until in the 1950s the FDA began processing reports that showed that low income southern state males, and dogs, were beginning to show sing of"feminization". In other words, DES or diethylstilbestrol traces in the chicken were building up in the tissues of 1st and 2nd generation males, altering male hormonal activity, resulting in un-descended testicles, higher pitched feminine range voices, abnormal breast enlargement, a small penis, impotence, infertility and increased risk for testicular cancer. What a price for a cheap meal! Chicken Says Goodbye To DES - Beef And Women Say Hello. Proving the brilliance of Big

Government"consumer protection", the FDA and Department of Agriculture created the necessary sort of media ballyhoo about chickens feminizing American males, with the result that DES or diethylstilbestrol was banned as a growth hormone in"chicken production".

However, in a left-hand-doesn't-know-what-the-right-hand-is-doing decision, the same governmental bodies sanctioned the use of DES or diethylstilbestrol for"beef cattle"production.

No surprise that beef cattle would be enhanced by DES, cattle would grow larger quicker, move to the slaughtering sheds almost a month earlier, cost less to feed and produce more profits, yet result in a continuing supply of low cost meat for the American consumer! Just like we like it!

And where were you in the"food chain"? If you had a steak in the years 1954 to around 1981, then you were munching on residue DES, antibiotics and who-knows-what-else. About 95% of total US beef production exploited the nature-bending properties of diethylstilbestrol!

What About Pregnant Women? By 1957 pharmaceutical ads run in obstetrics and gynecology publications trumpeted the claim that"you can build a bigger stronger baby by using DES"despite earlier known research that hormone manipulation during pregnancy was risky business. Who knows how many diethylstilbestrol scripts were handed out, like candy or vitamins to unsuspecting Mums?

For"DES girls", as these offspring were soon labeled, cervical cancer and higher incidence of irregular menstruation, structural malformation of the vagina and cervix, future miscarriages, premature births and infant deaths during their reproductive future. Crazily, the bad news that finally revealed that DES was a huge marketing sham also included news that it could be used as a"morning after"pill to terminate pregnancy!

World Views. No wonder that Europe banned US beef imports in 1989 due to fear of the drug-laced beef production habits of American producers."You don't want us...then we'll take our beef to other markets"came the instant reply from

American producers who let no grass grow beneath their feet when turning to Asian and South American markets. When the price is right, hormone free beef no longer is the issue.

HERB FARMING

You may want to set up a network with other herb farms to buy from, at wholesale prices. This way if you get low on some herb stock you have a place to obtain it for your customers. Remember, they are depending on you and you want to keep their business.

Have you noticed the high price of fresh herbs in the supermarket lately? It amazes me why more people aren't getting into the lucrative area of growing and processing fresh herbs. If you want to try your hand at herb farming, grow herbs that do well in your area. Start small, you can expand later with a green house, climate control and soil manipulation. You can begin with just a hot frame to start your seeds, and a screened area for drying your herbs away from insects and other undesirables.

Do a lot of homework before you start your herb farm. Go to the library or bookstore and find out what are the needs of the herbs you plan to plant. After you know what herbs you will be planting on your herb farm go through the history of that particular herb. Also see if it has any folklore or medicinal uses attached to it. Make up little booklets with this information as a"give aways"with your plants, it makes it more attractive to the buyer and will add to your sales and profits.

During your research, include study on which herbs attract pests. (Basil attracts white flies which are difficult to eliminate). Another trick is to keep herbs that attract pests away from the ones that do not on your herb farm.

If you are proficient with your computer, make up nice labels for your herbs. If not, go to a printer and have them make them for you. When you get to the stage that you are shipping herbs, always put a little information on the herb in the package. It gives your package a personal touch and your customer will be more apt to buy again.

Once you have a crop in you will need to market your

inventory. Go to Gourmet Restaurants and Gourmet stores. Don't forget health food stores and caterers. Talk to the chef or owner and explain what you have. You may try to leave some of your stock at the establishment on consignment to get your customers to see the value of fresh herbs. Offer to set up a display with a poster for their window. Or offer them a"special introductory price". You may even know some people in your area that just love to cook and would be ecstatic if they were part of your route. Exposure is key at this stage for both your name and your herbs. Go back to these potential customers often to build up your reliability.

Another important step is to write press releases that you can send to your local newspapers. Here is where the research on the folklore will come in handy. Use your gathered data on the folklore attached to herbs in the press release. These press releases will introduce you, and your products to the marketplace. Post them on the web as well, this will help you get customers from out of town. Be imaginative on the many interesting uses of fresh herbs and of course, how to find you as the source.

You may want to set up a network with other herb farms to buy from, at wholesale prices. This way if you get low on some herb stock you have a place to obtain it for your customers. Remember, they are depending on you and you want to keep their business.

USE ORGANIC FARMING

Finally, the commercialization of organic farming led to the weakening of regulations that permitted to use such inputs or products that were unimaginable a short time before. For example, in April 2004 USDA adopted a directive that permitted to use some synthetic pesticides on organic farms.

It is a widely known fact that many consumers prefer products of organic farming. Their choice may be explained by those advantages that organic farming has over conventional farming. Especially, when the market is overwhelmed by suspicious products and very often people even don't know what they really eat. So, advantages of organic farming are as follows: First of all it is traditionally supposed to produce natural food without artificial chemical additives that make such food much more attractive for a person who cares about his/her health because it is well known that pesticides often used in chemical farming may be very dangerous not only for human health but also for environment where people live. Organic farming traditionally uses manure, different composts and other natural products.

Secondly it fertilizes soil and in such a way makes it more convenient for farming. It is especially important because on using chemicals or other conventional means of farming the condition of soil may be seriously deteriorated. Nowadays it is already a problem how to restore the productivity of lands that were exhausted by pesticides and risk to be completely destroyed if such methods of farming won't be stopped. Also organic farming provides for diversity of plants while large farms where pesticides and other chemicals are used mainly have only one plant because it is more profitable for them as they don't need to waste money on using different artificial materials depending on a sort of plant. The nature of organic farming demands such diversity because it is needed for higher productivity of land.

One more positive aspect of organic farming is the fact that cattle and poultry on such farms are fed by natural food and the use of any chemicals such as antibiotics is highly restricted and their use is exceptional. Unlike conventional farming, organic farming treats them naturally. It means that they have natural conditions of life and they grow in normal conditions. It guarantees that the products of organic farms will be safe and healthy.

As for economical aspect, organic farming do not receive any financial help from the government unlike conventional farming which is subsidized and supported by means of various special terms of work. Though organic farming is still quite expansive, even if farmers do not buy synthetic pesticides and other chemicals for their farms.

Finally, though it seems a bit paradoxically but organic farming do not use some achievements of modern science.

Among them we can name genetically modified organisms. Though they are permitted to be grown on some farms in the USA the effect they produce on nature and human life are vague and not clear. This lack of knowledge about their effect makes products which are genetically modified unpredictable and potentially dangerous. All these facts contribute to the fast development of organic farming and growth of its popularity as well as the growth of interests from the part of the 'big' agribusiness. It makes organic farming very prospective branch of American agriculture that sooner or later could compete with conventional farming.

Contras of organic farming: But despite growing popularity of organic farming this way of farming has its own critics and, certainly, some of their arguments had to be thoroughly analysed. As I have already mentioned organic farming has enough advantages in comparison with chemical farming but some of this advantages are criticized and here I am going to concentrate my attention on this critic.

Firstly, it is necessary to underline that the growth of organic farming and the interest of agribusiness in it can have a negative effect on organic farming because it becomes more commercialized and the profit of the owner becomes the main aim of farming. As a result, nowadays some organic farms are allowed to use some inputs which, potentially, may have a negative impact on environment and human health (eg. sewage). That proves that further chase for profit may threaten to the positive image of organic farming and it also leads to the same problems that conventional farming has already got.

Among these problems I have already named the danger of pesticides used in chemical farming. But critics of organic farming estimate that some synthetic pesticides are made on the basis of natural pesticides. This improvement makes them less dangerous but not absolutely safe.

One more problem of organic farming is the protection of farms from pests. It is a well-known fact that conventional farms solve this problem using different chemicals that is not acceptable in organic farming. But in the case of wide spread of organic farming and its domination over conventional one the problem risks to become a real disaster if methods of natural anti-pests protection won't be as effective as their artificial counterparts. It is very difficult to foresee the prospects of organic farming in such a situation, especially, if we take into our consideration the fact that conventional farming and use of artificial means of plants protection forced pests to evaluate and to enforce their vitality and resistance.

Another argument of critics against organic farming is the danger of food contamination by products of this way of farming. Some specialists consider that organic food can be quite dangerous and provoke some diseases which are caused by food. They also say that some materials used in organic farming are dangerous too (manure contains a human pathogen, some biological toxins, etc.). But as for me I think that there are some other factors that can cause food contamination: the conditions of storage, the transportation of products, etc.

Finally, the commercialization of organic farming led to the weakening of regulations that permitted to use such inputs or products that were unimaginable a short time before. For example, in April 2004 USDA adopted a directive that permitted to use some synthetic pesticides on organic farms. In general such decisions and trends contradict to traditional principles of organic farming which emphasize the presence of only natural ingredients in organic products.

Thus, organic farming has some doubtful aspects that were listed in this part of my work but before criticize we have to study and compare both organic and conventional farming and nowadays the information concerning two ways of agriculture is obviously not enough, particularly, in relation to organic farming that is not so wide spread as its critics.

Summary

In my work I told enough about positive and negative aspects of organic farming but I still strongly believe that it is much more prospective than wider spread conventional farming. Though I have some doubts about its future because I think that it could hardly survive in its pure form but what I know for sure is that we must do our best to develop this branch of agriculture if we want to save our planet and our own health as well. For I think that any human activity must be as close to nature as possible because any intervention of human beings into natural processes very often has a negative impact on our environment and, consequently, on people themselves. That is why traditional organic farming with its main principles based on use of natural materials in agriculture is the most appropriate way of farming which permit people live in accordance with laws and principles of nature.

THE FARMER

Think of a pregnant woman. Once fertilization happens, and she's conceived, even though it takes months for any evidence of all that activity to become visible, you know that there's a miracle taking place the whole time. And if all goes well in that pregnancy, she knows to expect a healthy baby at the end of 9 months (her gestation period).

When a farmer plants a crop, *as long as the soil is fertile*, he knows to *expect* a harvest in good time.

He knows to expect multiples of what he's planted.

He knows to patiently nurture the soil, keeping it watered and nourished while he waits to reap what he's sown.

Notice that he doesn't agitate over how soon his harvest shows up. He knows it's going to show up in good time, as long as the environment remains conducive. He doesn't worry about 'if' it's going to show up. He *knows* it will.

How does that apply to you?

Well let me ask you: have you ever felt discouraged because things weren't happening as fast as you'd like? Perhaps you're doing all the things you should: you've written your goals, made plans that you're following, faithfully affirming and visualizing daily, practicing daily gratitude, etc.

Still nothing seems to be happening, and you're starting to wonder if you should quit. Perhaps you're starting to wonder if your goal was worth going for after all. You've probably told yourself that the goal was too big for you anyway.

If this is you, listen: learn from the farmer. You don't know exactly when your harvest will manifest, but realise one thing: the moment you put that goal on paper and started working towards it, you set a chain of activity in motion - activity which will end up as your goal becoming real.

Think about it. When a farmer plants a seed, he doesn't see the shoot the next day does he? Yet the moment that seed is in the soil, the growth and manifestation process begins. Right at that instant. This initial phase is invisible to us on the outside but that doesn't change the fact that it's happening.

When you set a goal, and you've written it down and are focusing on it, *know* that (as long as you satisfy the right conditions), your desires will take on physical form in good time. The moment you decide on something, the moment you think it, you set in motion activities geared towards bringing that thought to form.

Your idea/desire/goal, is a 'seed' from your conscious mind. Your subconscious mind is the 'soil' in which you plant this seed. By natural law, the moment you plant a seed in soil, everything necessary for that seed to bring forth multiple fruit starts happening. That's all soil knows to do - grow stuff.

Your subconscious mind works the same way: all it knows to do is manifest stuff - it brings about the physical manifestation of any thought you plant in it. Note that it doesn't matter what you plant in it (read:"it doesn't matter what you ask for"or"it doesn't matter how big or small your goals are"- as long as you work in harmony with Universal laws, they will become physically real).

This physical manifestation does *not* happen overnight, however. That's what I want you to remember: as long as you're doing what you know to do, and you're staying focused on your desired outcome, rest assured that at the appropriate time, your vision *will* be made real.

That is what the law of Gestation (one of the Universal Laws) is all about: it refers to the period between the planting of a seed and the harvest/birth/end result.

According to this Law, there is a period of time that must elapse before that thought will manifest in form. The length of that time is dependent on the magnitude and intensity of the energy of the thought.

When you understand this Law, you'll find it easy to go about your business with full faith, expectation and gratitude that your vision is daily becoming real.

Think of a pregnant woman. Once fertilization happens, and she's conceived, even though it takes months for any evidence of all that activity to become visible, you know that there's a miracle taking place the whole time. And if all goes well in that pregnancy, she knows to expect a healthy baby at the end of 9 months (her gestation period).

Your goals have gestation periods too.

Don't give up prematurely. *Persist* in doing the right things, fully *expecting* your desired result, and fully *grateful* for that result being already present.

By law (as you work in obedience to Law), it has to happen. Your goals and desires must manifest, for the Universal laws are precise. They never fail.

So anytime you are tempted to get discouraged at the way things are going, stop and remember the farmer (and the pregnant woman:). Then get up and keep going forward until you reach your goal.

Chapter 6

Genomics and Proteomics

PROTEOMIC STUDIES OF STEM CELLS

Stem cells of both embryonic and adult origins hold great promise in regenerative medicine owing to their unique properties of unlimited self renewal and differentiation towards specific lineage(s) once they receive the proper signals. Proteomics is a series of technology platforms driven by advancements in mass spectrometry and bioinformatics that encompass protein identification, the relative quantitation of proteins and peptides, their subcellular localization, and studies of post-translational modifications and protein-protein interactions. Stem cell biology has been influenced by these approaches and has evolved in the post-genomics era. Among many challenges in stem cell biology, there is a pressing need for the implementation of proteomic applications.

Recent work on stem cells using proteomics has shown that transcriptome analyses fail to provide a full guide to developmental change in stem cells, and protein interactions that can only be discovered systematically using proteomic approaches have yielded important new concepts on processes regulating development and stem cell pluripotency. In this, we will review current proteomic studies on embryonic and adult stem cells with an emphasis on embryonic stem cells.

INTRODUCTION

Stem Cells and Proteomics

Stem cells of any type are defined by two distinct

properties. The first is indefinite self-renewal, a feature that provides for maintenance in a tissue and/or organism for an extended period of time. The second is the ability to differentiate into a number of different daughter cell types, unlike non-stem cells that are committed to a single lineage. Adult somatic stem cells are found in the majority of organs and tissues in adult organisms, and are thought to function in long-term tissue maintenance and/or repair.

In contrast, embryonic stem cells (ESCs) are derived from embryos and are unique in their ability to be maintained in vitro in a pluripotent state, i.e., capable of recapitulating all three germ layers and an entire organism.

Both adult and embryonic stem cells have provided distinct challenges for analysis. Adult stem cells tend to be rare, difficult to purify or maintain in culture. For this reason, adult stem cells have provided a greater technical challenge for large-scale transcriptome and proteome analyses. ESCs, by contrast, are readily grown to large numbers in culture, and have been utilized for extensive analysis by transcriptional profiling and other genome wide techniques. In addition, ESCs are easily manipulated in vitro, making them ideal tools for probing stem cells properties and characteristics using a wide variety of techniques.

In the current post-genomic era, in which transcriptome mapping using DNA microarray technology is commonplace, consideration of the transcriptome alone offers an incomplete and biased interpretation of the underlying stem cell biology. Inherent problems associated with such a transcriptional profiling approach include first, the analysis is obviously limited to genes present on the microarray, and it is possible that there are"stemness"genes that have not yet been identified and are not represented in the chips used; second, changes at the mRNA level may not be proportional to changes in protein expression; third, protein complex formation, numerous post-translational modifications (PTMs) and protein degradation greatly impact protein-protein and protein-DNA interactions, making the functional output of these systems virtually impossible to predict based solely upon gene expression and/

or genomic data. The term"proteome"was originally coined by Wilkins et al. to describe the total set of proteins expressed in a given population, a.k.a. cell, tissue, organelle, organism, or pathological state. The term"proteomics"refers to a set of techniques well suited to identify proteomes, but has been broadened to include large-scale techniques capable of identifying proteins, and analyzing both their structures and their functions at a genome wide level.

Proteomics encompasses a wide variety of techniques, ranging from yeast two-hybrid screens for identifying protein-protein interactions, antibody-based protein chips for identifying proteins and high throughput crystallography screens to provide structural analysis. All of these techniques provide invaluable insights into the proteome and its function in a cell, but perhaps the most widely utilized group of techniques centre around mass spectrometry (MS), which will be further discussed below (also reviewed in Aebersold and Mann and Cravatt et al.

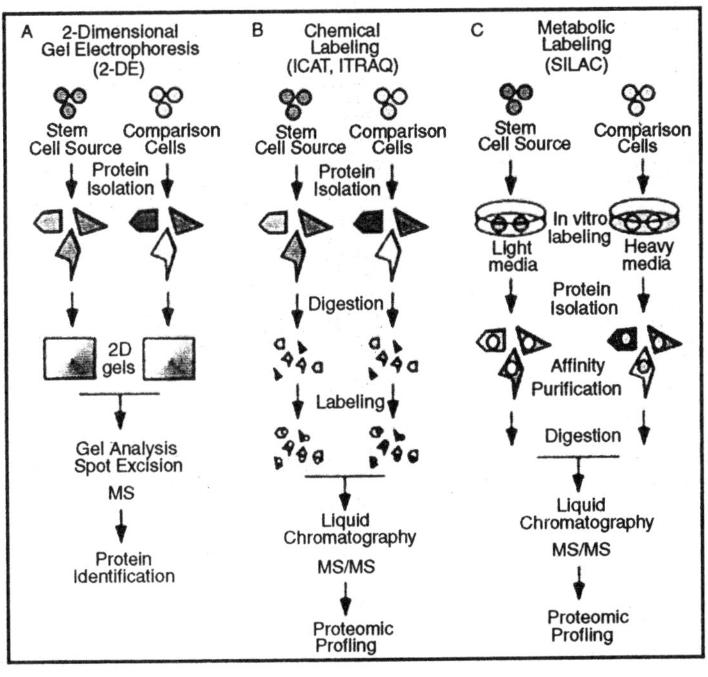

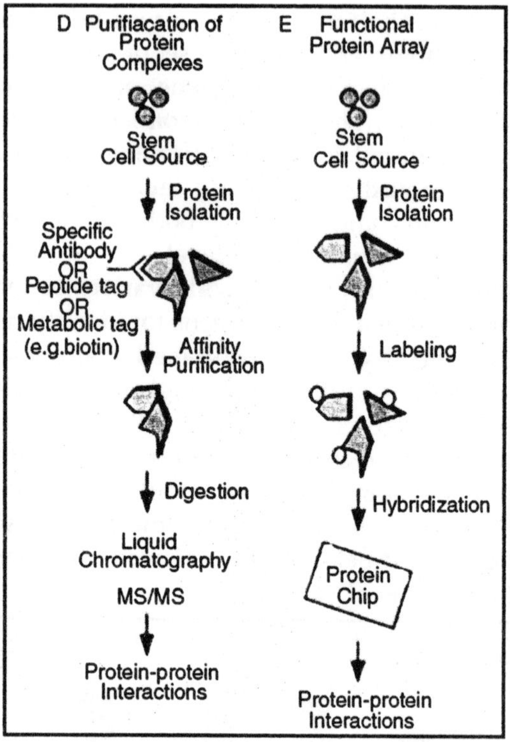

Summary of Proteomic Approaches Utilized in the Study of Stem Cells

Advantages and disadvantages of the various methods that have been used in proteomic profiling and protein interactome mapping are highlighted. A. Two dimensional electrophoresis (2-DE) followed by mass spectrometry (MS) analysis has been widely used to compare various populations of stem cells to more differentiated cell types. B.iTRAQ (isobaric tags for relative and absolute quantification) and ICAT (isotope coded affinity tags) are two chemical labeling approaches that have been used prior to tandem mass spectrometry (MS/MS) analysis in comparison of purified stem cell populations. C. SILAC (stable isotope labeling with amino acids in culture) is a metabolic labeling approach that has been used in proteomic profiling and quantitative phosphoproteomic studies. D. Affinity-based purification of protein complexes followed by MS/MS analysis has been widely used in protein interactome studies. E.The use of functional protein arrays is a very promising approach that has predominantly been utilized in yeast protein interactome studies.

Mass Spectrometry (MS)-based Platforms for Proteomic Research

MS functions by ionizing relatively small molecules and then measuring their mass to charge ratio (m/z). While traditional MS itself is capable of identifying the mass of a highly purified small molecules, it can do little else with more complicated molecules (such as peptides) or mixtures of samples. To further increase the range of substances that can be identified by MS, two can be combined in tandem (termed MS/MS) in which a peptide first has its molecular mass measured (in MS1) and then is bombarded with electro-neutral gases to cause fragmentation. The m/z ratios of these resulting smaller fragments are then measured in the second analyser (MS2), and following computer analysis the amino acid sequence of the peptide can be determined.

Thus, virtually any single peptide, in a relatively purified state can be identified using tandem MS. For the analysis of a whole protein, or even multiple proteins, further manipulations are needed to purify the mixture to reduce the complexity of any specific input into the MS. This is usually accomplished first by purifying a sample by simple SDS electrophoresis and subsequent excision of relevant band(s) or a whole lane of a gel which is then broken down into smaller fragments. These gel fragments are then digested in situ with a protease (typically trypsin) and the peptides are recovered.

To further fractionate the specimen prior to analysis by MS, the samples undergo either single dimensional liquid chromatography (LC, typically a reverse-phase LC which separates based upon hydrophobicity), or multidimensional (LC/LC), with the choice based upon the complexity of the initial sample. Subsequent to LC or LC/LC but prior to application to the mass analyser, the peptides are ionized, usually by electrospray ionization in which a potential is applied across a fine needle through which passes the elute from the LC column, creating a fine spray that forms droplets containing the sample, and heat applied prior to entry into the MS allows for desolvation and ionization. MS/MS analysis ensures identification of the peptide.size and amino acid sequence.

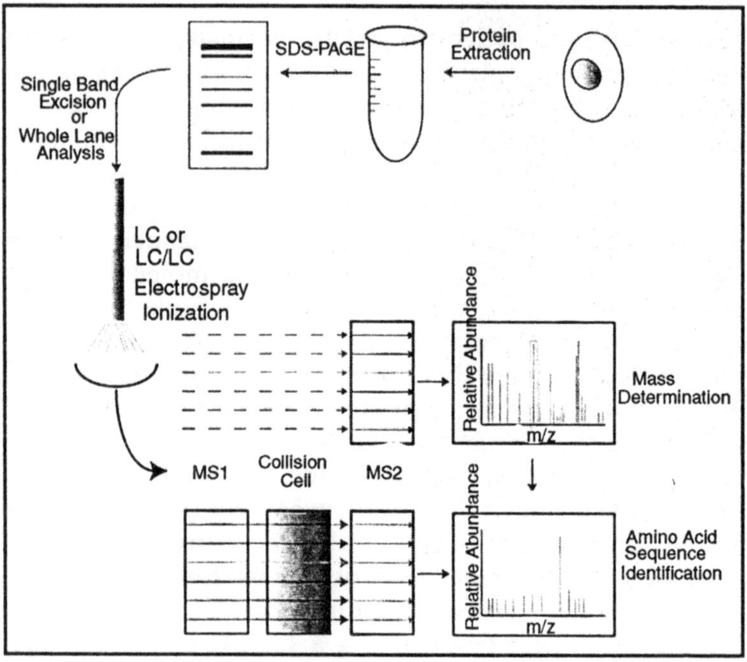

Schematic Depiction of LC-MS/MS Procedure

Protein extracts are made from stem cells of any type and first fractionated by SDS. Individual bands or the whole lane are then subjected to in situ digestion with trypsin. After trypsin digestion, the product is subjected to either single-dimensional (LC) or multi-dimensional (LC/LC) liquid chromatography for further separation of the mixture.

Elute from LC is then ionized by electrospray ionization, and each elute peak first passes through an MS (typically MS2) for mass determination. Each peak is then separated in MS1 and passes into a collision chamber where it is further fragmented and subsequently analysed in MS2, which aids in peptide sequence determination. After completion of this process, a complicated protein mixture is reduced to a fragment ion spectrum and molecular weight for each peptide. Bioinformatics is then necessary to translate each specific spectrum into a peptide and protein from which it originally arose. The algorithms involved are varied and complex, but are based upon comparison to the theoretical spectrum of known proteins from a database and de novo sequencing in which each fragment spectrum is directly translated into a specific peptide, or a hybrid approach that combines both.

Each peptide is then mapped to a protein based upon either a deterministic (i.e., a predetermined algorithm such as in or probabilities of a match. The result is an identification of all possible protein(s) in a given sample. The use of LC coupled tandem MS/MS has allowed for two general approaches. The first is termed"shot-gun"proteomics, in which a single sample, such as a cell line, tissue, or highly purified cell population is analysed to assess all peptides/proteins expressed.

This is also known as expression-based proteomics. The second is affinity purification, in which a single protein species is purified from a cell; with the goal being to identify associated molecules. While both methods have been widely utilized, affinity purification has provided unique insights into network properties of organisms and stem cells, and thus often been referred to as functional proteomics. In general, affinity purification is based upon two techniques. First, affinity purification can be performed on native proteins using antibodies to isolate a single protein and its associated proteins.

The major drawback is that the antibody can often be the limiting reagent, making it difficult to purify rare proteins or large amount of complexes. The second involves attaching a specific peptide tag to a cDNA of interest, allowing for easy purification and elution of the tagged protein of interest. These methods also typically allow for eluting the affinity tagged complexes from the column by proteolytic cleavage at a specific recognition sequence. A variation of this tag-based technique is based upon metabolic tagging with biotin. Cells are generated which express the E.Coli derived BirA ligase capable of attaching biotin to a specific peptide recognition sequence.

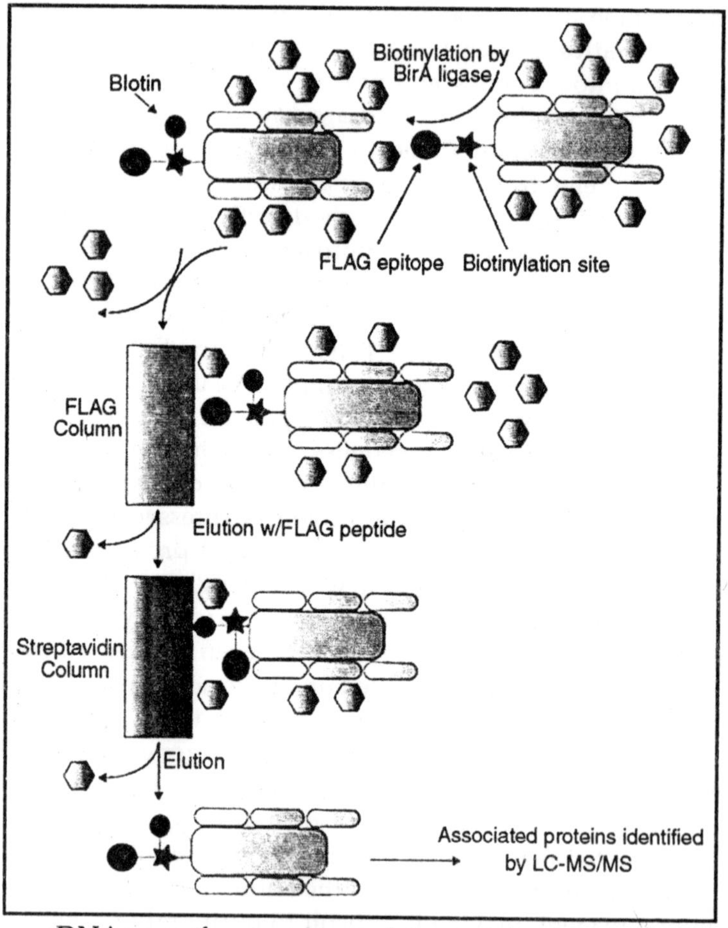

cDNAs are then engineered to contain the recognition sequence, allowing them to be efficiently biotinylated in vivo and captured in vitro due to the strong affinity of biotin for streptavidin. The predominate advantage of metabolic tagging methods is the exceptionally high affinity of streptavidin for biotin ($K_d \approx 10^{-15}$, as opposed to a $K_d \approx 10^{-9}$ for calmodulin binding protein).

Using either tagging approach, the tags are often combined to allow for tandem purification, thereby increasing the purity of the complex and the specificity of the subsequently identified interactions.

There are several advantages associated with this affinity purification-MS method: first, it can be performed under relatively physiological conditions; second, it does not typically perturb relevant PTMs, which are often crucial for the organization and/or activity of complexes and can also be identified by MS; third, it can be used to probe dynamic changes in the composition of protein complexes when used in combination with quantitative proteomics techniques such as iTRAQ and SILAC.

Schematic Depiction of two Affinity Purification Approaches

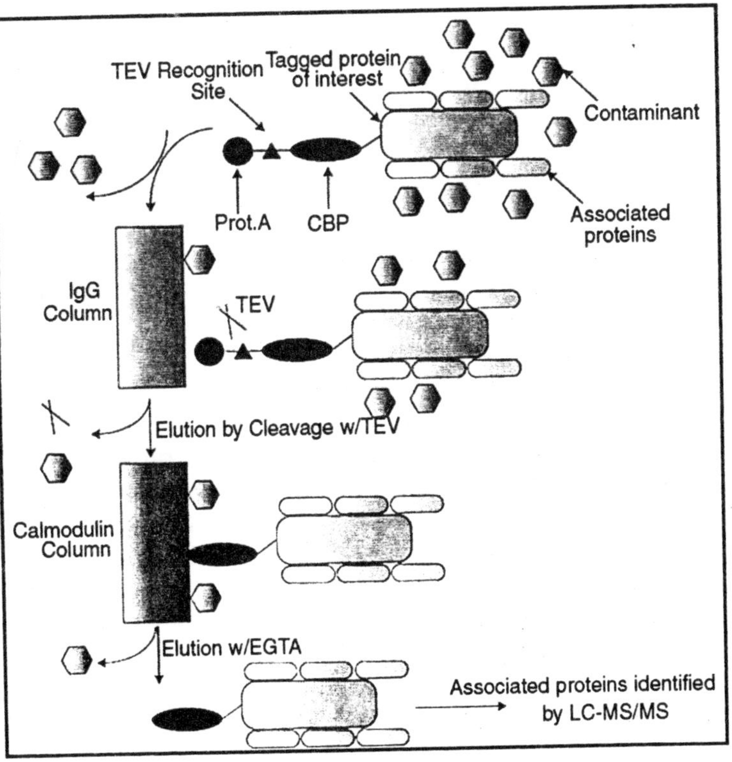

A. Tandem affinity chromatography: a protein of interest is first engineered to contain a Protein A (filled in circle) tag, a Tobacco Echo Virus recognition site (TEV, filled in triangle), and a calmodulin binding peptide (CBP) tag (filled in ellipse). Extracts are made from cells expressing the tagged protein, which should contain associated proteins and contaminants. These complexes are bound to an IgG column, and washed to remove majority of contaminants.

TEV protease is then used to elute the semi-purified protein complexes which are subsequently absorbed onto a calmodulin column. After further washing, purified protein complexes containing the protein of interest and its associated proteins are eluted by calcium chelation (EGTA) and identified using LC-MS/MS.

Schematic Depiction of two Affinity Purification Approaches

B. Metabolic tagging for affinity chromatography: First, stem cells are engineered to express the E.Coli derived biotin-ligase BirA, which attaches biotin to a defined recognition sequence, shown with a filled-in star. Proteins of interest are then engineered with a biotinylation site and a FLAG epitope. Inside the stem cell, the biotin (blue circle) is added by BirA. Protein extracts are made and applied to a FLAG-antibody column, and after washing semi-purified complexes are eluted with FLAG peptide. The elute is then applied to a streptavidin column, and after washing the purified protein of interest and its associated proteins are eluted by denaturing and identified by LC-MS/MS.

In addition to identifying large arrays of proteins as well as protein complexes, proteomics has also advanced to be more quantitative, i.e., allow for protein levels to be directly compared between two samples. While there are a number of techniques, two are most widely used: ICAT and iTRAQ. Briefly, proteins from two populations of cells are labeled using different chemicals with different isotope compositions (i.e., hydrogen vs. deuterium in the case of ICAT or an analogous four isotope tag in iTRAQ), and the samples are then remixed and quantitative protein levels can be assessed. The advantages of these techniques are that both allow for the quantitation of virtually any sample, and very large samples are possible, although issues with labeling efficiency and over-labeling can cause difficulties.

In contrast, SILAC uses a similar approach in which two populations of cells are labeled with isotopically distinct amino acids in vivo and then analysed, allowing for differences between the two cell populations to be assessed. The advantage of this technique is that labeling efficiency and over-labeling are no longer an issue, although it is a difficult procedure to scale up to larger, proteome scale procedures. The development of these MS-based technique platforms has greatly advanced the proteomic studies of stem cells, which is discussed in detail in the next two sections.

PROTEOMIC STUDIES OF EMBRYONIC STEM CELLS (ESCS)

The ESC Proteome

Since their discovery over 25 years ago, murine embryonic stem cells (mESCs) have provided an invaluable tool for answering genetic questions. With the establishment of human embryonic stem cells, new opportunities for tissue repair or replacement are being actively explored. To complement the transcriptomic analyses of ESCs that define a genome wide RNA expression signature of stemness, stem cell proteomics provides an excellent tool to characterize ESCs at protein level and derive a protein pluripotency signature that may disclose novel ESC-specific benchmarks.

The proteomic analysis of embryonic stemness has been probed using MS-based protein profiling of both undifferentiated and differentiated ESCs. A quest for human (line HES-2) and mouse (line D3) ESC-specific proteins resulted in 1,775 non-redundant proteins in hESCs, 1,532 in differentiated hESCs, 1,871 in mESCs, and 1,552 in differentiated mESCs with a false positive rate of <0.2%. Comparison of the data sets distinguished 191 proteins exclusively identified in both human and mouse ESCs, many of which are uncharacterized proteins and are potential novel ESC-specific markers or functional proteins Elliott et al. utilized 2D gels with multiple pH gradients and varied acrylamide concentrations to resolve approximately 600 protein spots from mouse R1 ESCs on silver stained gels and represents the initial step in producing a comprehensive ESC 2D protein database. Nagano et al. using an automated microscale 2D LC-MS/MS analysed total proteins in mouse E14-1 ESCs. They assembled a catalogue consisting of 1800 proteins, containing many components derived from ESC-specific and stemness genes defined by the transcriptome analysis, and a number of components, such as Oct4 andUTF1, which are expressed specifically in ESCs. Importantly, they detected ESC-specific transcription factors of low abundance (10^4 to 10^5 copies/cell) and found 36% of total proteins were located in the nucleus, consistent with the high nuclear to cytoplasmic ratio of ESC colonies.

Recently, Graumann et al. fractionated the SILAC-labeled ESC proteome by 1D/IEF (isoelectric focusing) followed by high resolution analysis on a linear ion trap-orbitrap instrument (LTQ-Orbitrap) to sub-ppm mass accuracy which resulted in confident identification and quantitation of more than 5,000 distinct proteins. This is the largest quantified proteome reported to date and contains prominent stem cell markers, such as Oct4, Nanog, Sox2, Utf1 and an embryonic version of Ras. Bioinformatics analysis of the ESC proteome reveals a broad distribution of cellular functions with overrepresentation of proteins involved in proliferation. In addition, Graumann et al. compared the proteome with a recently published map of chromatin states of promoters in ESCs and find excellent correlation between protein expression and the presence of active and repressive chromatin marks.

An interesting feature of the ESC proteome in the Nagano study and another study in D3 ESCs is that it retains the cell surface markers and signaling molecules that are characteristic of differentiated cells. This is not inconsistent with the notion that interactions between cell surface proteins and extracellular ligands are key to initiating ESC differentiation to specific lineage. Although it is formerly possible that a small portion of cells were differentiated to a variety of cell lineages during the culturing condition, it is tempting to hypothesize that the ESC proteome is equipped with multiple protein components unique to a number of differentiated cell types, enabling cells to respond to various external signals leading to differentiation to specific lineages, a property of pluripotency of the ESCs. So far, relatively little is understood regarding how stem cells are programmed towards a particular cell lineage. This is an important area of investigation that involves directed differentiation to influence the lineage commitment of these pluripotent cells in vitro. Manipulation of extracellular signals and overexpression of transcription factors can drive ESCs to commit to a specific cell type, however, ultimately it is the changes in nuclear expression that direct differentiation down to a specific lineage. Accordingly, nuclear proteomics-studies of collective actions and interactions of proteins found in the nucleus-has been proposed to inventory nuclear proteins in both undifferentiated and differentiating cells and decipher their dynamics during cellular phenotypic commitment.

This provides an opportunity to identify unknown transcription factors and additional nuclear effectors critical in the maintenance of cellular phenotype. In addition, it offers insights as to what nuclear profile is needed to programme or reprogramme cellular fate with limited imprinting side effects.

The ESC Epiproteome

To identify biologically relevant proteins important for stem cell self renewal and pluripotency, the extensive catalogue and benchmark of protein databases are not sufficient. Many biochemical pathways are directed by changes in PTMs such as phosphorylation rather than by changes in abundance of proteins themselves. Studies have now shown that epigenetic mechanisms, such as covalent modifications of histones and DNA methylation are vitally important to the pluripotent nature of ESCs and that these mechanisms also regulate differentiation The epigenetic nature of the ESCs (the

ESC"epigenome") has been demonstrated to be unique and its characteristics have been strongly linked to the global permissivity of gene expression and pluripotency In analogy to epigenome, a new term"epiproteome"has been coined to reflect a protein landscape of PTMs and histone variants.

Phosphorylation is a critical PTM involved in modulating protein function. To gain insight into intracellular signals governing ESC self-renewal and differentiation, a multivariate systems analysis of proteomic data generated from combinatorial stimulation of mESCs (line CCE) by fibronectin, laminin, LIF was performed Phosphorylation states of 31 intracellular signaling network components were obtained across 16 different stimulus conditions at three time points by quantitative Western blotting, and computer modeling was used to determine which components were most strongly correlated with cell proliferation and differentiation rate constants obtained from measurements of expression levels. The study identified a set of signaling network components most critically associated with differentiation, proliferation of undifferentiated as well as differentiated cells.

A large-scale proteomic analysis of hESCs (BG01 and BG03 lines) was also performed using PowerBlot and Kinexus Western blot assays coupled with immunofluorescence. The study identified over 600 proteins expressed in undifferentiated hESCs, including a number of potential new stem cell markers, and highlighted over 40 potential protein isoforms and/or PTMs including 22 phosphorylation events in cell signaling molecules. More recently, a nucleosome-ELISA method was developed to assess quantitatively the status of PTMs and histone variants (dubbed"epiproteomic signature") present within the total cellular nucleosome pool.

The results indicate that assessment of the steady-state levels of PTMs and macroH2A yields an epiproteomic signature that can distinguish between ESCs, EC cells and MEFs. Furthermore, epiproteomic nucleosome signatures change in response to exposure of cells to small molecules such as RA and TSA and over the course of ESC differentiation. This indicates that the epiproteomic signatures are useful for investigation of stem cell differentiation, chromatin function, cellular identity and epigenetic responses to pharmacologic agents.

The direct analysis of a large number of peptides using 2D LC-MS/MS permitted the systematic identification of peptides carrying PTMs. Nagano et al. identified protein PTMs in a number of ESC proteins including five Lys acetylation sites and a single phosphorylation site. Phosphorproteome analysis of undifferentiated and differentiated mESCs using phosphoprotein affinity purification followed by 2D LC-MS/MS indicated that many chromatin-remodeling proteins are potentially regulated by phosphorylation. Interestingly, affymetrix microarray analysis indicated that gene expression levels of these sample proteins had minimal variability between the compared samples. These findings collectively highlight the critical roles that epigenetic factors play in maintaining pluripotency of ESCs, and stress the necessity and value of proteomic analysis.

THE ESC PROTEIN INTERACTION NETWORK

The expression-based studies of ESC proteome and epiproteome provide a comprehensive inventory of proteins as well as their PTMs, some of which may be used as ESC markers. However, such protein lists are not sufficient to describe biological processes. Vital cellular functions require the coordinated action of a large number of proteins that are assembled into an array of multiprotein complexes of distinct composition and structure. The analysis of protein complexes and intricate protein-protein interaction networks is a key to understanding virtually any complex biological systems including stem cells.

To understand how pluripotency is programmemed and maintained in ESCs, we have utilized a proteomic approach to isolate protein complexes and constructed a protein interaction network surrounding the pluripotency factor Nanog. The approach takes advantage of the extraordinary affinity of streptavidin for biotin, and obviates reliance on antibodies of inherently lower affinity for purification. It has been reported that single-step streptavidin capture of tagged transcription factors is sufficient to isolate specifically associated proteins with minimal non-specific contamination. In this system, BirA expressing ESCs serve as a recipient for other tagged cDNAs. A construct bearing the pluripotency factor with a FLAG tag as well as a peptide tag that serves as a substrate for in vivo biotinylation was expressed in ESCs. The tagged protein was recovered from nuclear extracts with streptavidin beads together with its potential interacting partners. For tandem purification, the nuclear extracts were first subjected to immunoprecipitation with anti-FLAG antibodies and the recovered protein complexes were further purified by streptavidin beads. Protein complexes recovered from either one-step streptavidin or tandem purification were subjected to microsequencing by LC-MS/MS.

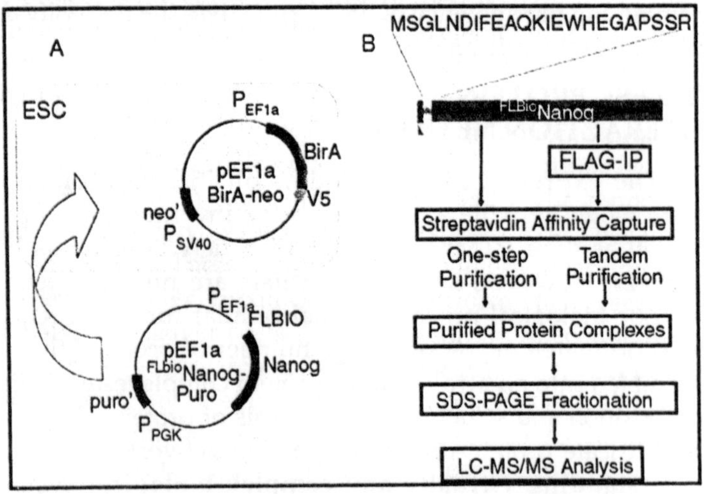

Fig. Strategy for Affinity Purification of Nanog Associated Protein Complexes in mESCs.

A. Establishment of a biotinylation system in ESCs. A stable ESC line expressing the bacterial BirA enzyme was first established by transfection with a BirA-expressing plasmid bearing the neomycin resistance (neor) gene and G418 selection; A second plasmid containing Nanog cDNA with an

N-terminal Flag-biotin dual tag (FLBIO) and a puromycin resistance (puror) gene was introduced and cells selected with puromycin. The resulting stable lines are resistant to both G418 and puromycin and express FLAG-tagged, biotinylated Nanog that can be immunoprecipitated by anti-FLAG and streptavidin antibodies/beads. B. Two complementary affinity purification strategies for protein compexes purification. Single streptavidin immunoprecipitation and tandem affinity purification (anti-Flag immunoprecipitation followed by streptavidin pulldown) were performed in parallel, the purified protein complexes were fractionated on SDS, and subjected to LC-MS/MS to identify components of the protein complexes.

We first chose to focus on the variant homeobox Nanog protein, considering its role in maintaining pluripotent state of cells in early mouse embryo and promoting pluripotency of mESCs. By affinity purification of Nanog associated protein complexes followed by LC-MS/MS, components of Nanog protein complexes (and thus direct and/or indirect Nanog-interacting partners) were identified. Many of the candidates identified were other transcription factors or components of transcriptional complexes, some of which had already been associated with ESC functions in previous studies. A number of novel and known critical factors were validated, both physically and functionally, for association with the bait Nanog and were used (together with another well known ESC marker Rex1) for purification of a second tier of complexes.

This iterative,"bottom-up"strategy reveals a tight, highly interconnected protein network greatly enriched in nuclear factors individually required for maintenance of ESC properties and co-regulated on ESC differentiation. In addition, the network links to multiple corepressor pathways, which provides both a means to regulate different sets of target genes and a fail-safe mechanism to prevent differentiation towards different lineages, a requisite for pluripotency. Furthermore, downstream gene targets of several core pluripotency factors identified from previous studies also serve as upstream regulators in the network, indicating that the ESC interaction network is a self-contained, exceedingly tight cellular module dedicated to pluirpotency.

Finally, identification of a number of network proteins that are not strictly specific to ESCs and cannot be identified by transcriptional profiling, highlights the importance and advantage of proteomic studies in ESCs.

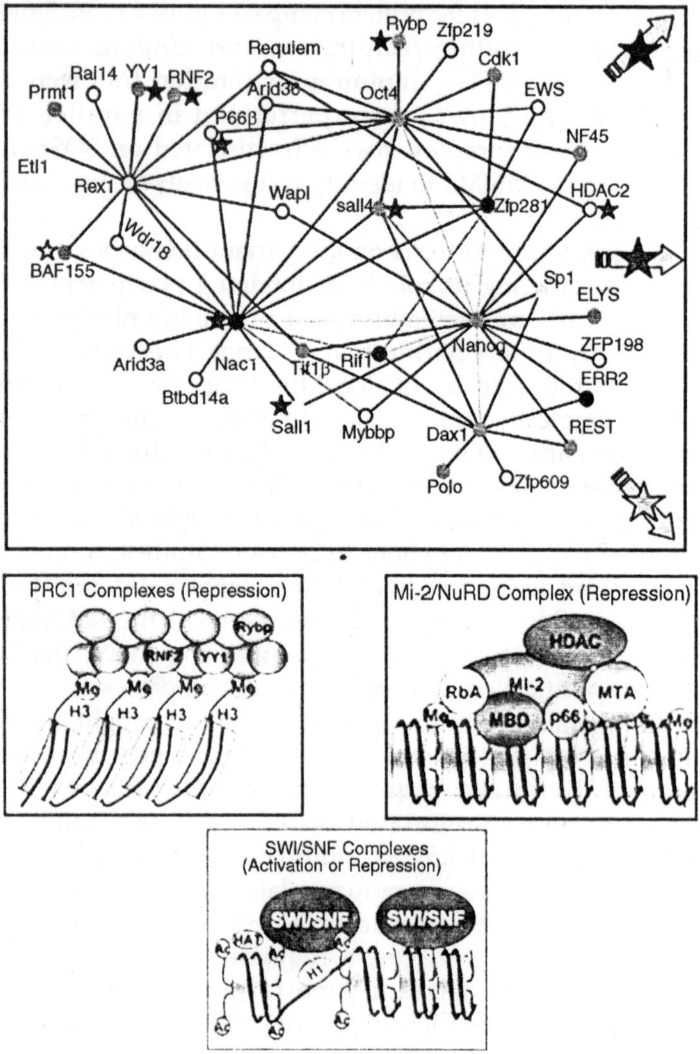

Fig. A Protein Interaction Network in mESCs.

A. Proteins with red labels are tagged baits for affinity purification. Green and red lines indicate confirmed interactions by coimmunoprecipitation or published data. Dotted lines indicate potential association. Green circles indicate proteins whose knockout results in defects in proliferation and/or survival of the inner cell mass or other aspects of early development; Blue circles indicate proteins whose reduction by RNAi (or shRNA) results in defects in self-renewal and/or differentiation of ESCs; Yellow circles are proteins whose knockout results in later developmental defects; White circles denote proteins for which no loss-of-function data are available. Also indicated within the network are three major chromatin modifying complexes whose components are marked with black stars (Polycomb repression complex 1), red stars (NuRD complex) and a blue star (SWI/SNF complex), respectively.

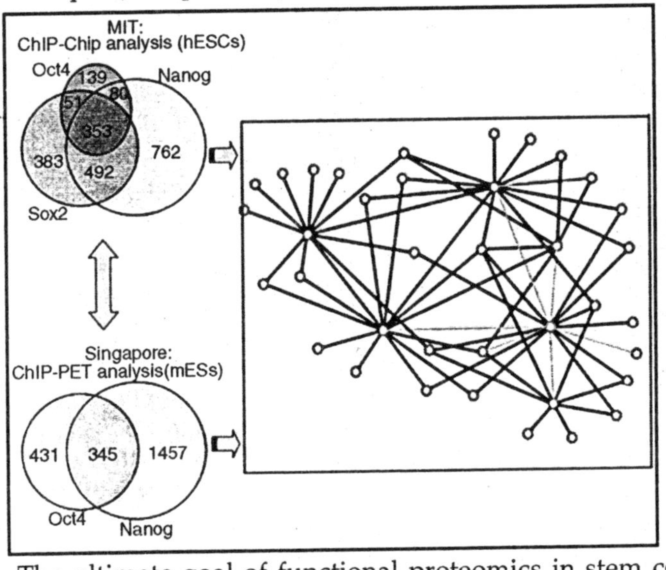

The ultimate goal of functional proteomics in stem cells is to decipher the molecular function of an entire cell by generating a construction master plan describing all molecular machines, their functions in maintain stem cell properties, their reactions to external stimuli during differentiation, and their interconnectivities. Our work on the protein interaction network in mESCs described above represents the first step towards that direction. In addition, it provides a framework for exploring the combinations of factors that may permit optimal reprogramming of differentiated cells to an ES cell state.

The ESC Transcriptional Regulatory Network

Large-scale transcriptomic and proteomic analyses of ESCs are complementary to each other and have laid a foundation for a better understanding of the underlying stem cell biology.

However, missing links exist such as gene transcription may not directly be indicative of or proportional to protein translational readout (expression), and conversely, protein expression and multiprotein complexes do not themselves specify target gene regulation of the protein(s). A comprehensive understanding of establishment of the pluripotent state in ESCs requires construction of an expanded transcriptional regulatory network in which many key transcription factors besides Nanog, Oct4 and Sox2 and their interaction partners bind directly to their target genes.

Recent studies have begun to elucidate transcription networks surrounding the three core ESC transcriptional factors Nanog, Oct4, and Sox2 that operate to control ESC pluripotency. Using ChIP-chip analysis (chromatin immunoprecipitation followed by microarray hybridization to identify binding sites on a genome wide scale), Boyer et al. showed that Oct4, Sox2 and Nanog collaborate to regulate hESC pluripotency and self-renewal through autoregulatory and feedforward loops.

These three transcription factors function by activating pluripotency genes including themselves and by repressing key developmental genes possibly in part with aid of Polycomb proteins. Using ChIP followed by paired-end ditags (ChIP-PET) approach, Loh et al. surveyed target genes of Nanog and Oct4 in mESCs and found that both regulate substantially overlapping target genes. However, cross-examination of the target genes of Nanog and Oct4 between hESCs and mESCs revealed a limited overlap between the two sets of data, suggesting either different control mechanisms between the two species or inherent variations between the two technique platforms. The result emerged from these studies was the high degree of overlap between the genes targeted by pairs or all the three transcription factors. However, questions remained to be address as how other factors besides the three in the protein interaction network contribute to maintenance of stem cell identity and how the multiprotein complexes specify target gene regulation.

Although neither expression nor transcription factor binding studies in isolation are sufficient to establish a regulatory relationship between a transcription factor and its targets, integrating these methodologies has provided two independent sources of evidence for high confidence prediction of novel transcriptional networks regulating ESC self-renewal and commitment.

Using a modified ChIP-chip procedure combined with affinity purification and LC-MS/MS to expand the current protein interaction network, Kim et al. systematically surveyed target genes of total 9 protein interaction network factors c- and constructed an expanded transcriptional regulatory network in mESCs.

This network contains many more core pluripotency factors in addition to Nanog, Sox2 and Oct4that form autoregulatory and feedforward regulatory circuitries. In particular, Klf4 serves as an upstream regulator of larger feedforward loops containing Nanog, Sox2 and Oct4 as well as c-Myc. More importantly, combined analyses of bioChIP-chip data with gene expression data revealed that majority of common targets of over 4 factors are highly active in ESCs and repressed upon differentiation.

In the case of targets bound by fewer factors, both active and repressed genes are present and the balance shifts towards gene inactivity with reduced factor co-occupancy. The extreme is that distinct targets of a single factor are largely inactive or repressed. Moreover, the regulatory network also indicates that c-Myc and three other factors play distinct roles in ESCs, i.e., c-Myc is largely involved in stimulation of cell proliferation and regulation of chromosomal accessibility; whereas Oct4/ Sox2/Nanogpositively regulate ESC factors and negatively regulate differentiation (Kim et al., 2008). This provides a potential mechanism that might account for the differential regulation of transcription factor targets in ESCs and provides mechanistic insights into the 4-factor mediated somatic cell reprogramming.

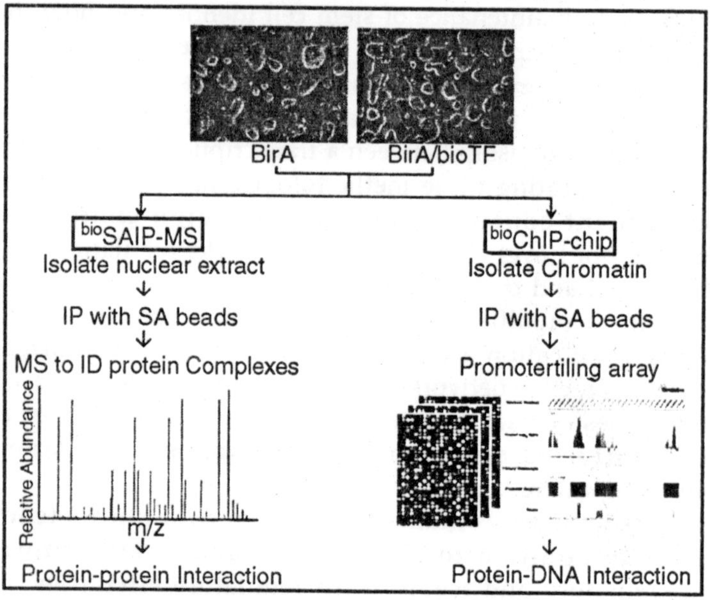

Fig. Strategies for Mapping Protein-protein and Protein-DNA Interactions in mESCs.

The ESCs expressing BirA alone (as control) and BirA plus biotinylated transcription factors (bioTF) can be used for isolation of protein complexes using streptavidin (SA) immunoprecipitation (IP) coupled with LC-MS/MS (dubbedbioSAIP-MS) and construction of a protein-protein interaction network; meanwhile, the same ESCs can be subjected to in vivo biotinylation-mediated chromatin immunoprecipitation and microarray (dubbed bioChIP-chip) to identify protein-DNA interactions and construct a transcriptional regulatory network.

Our demonstration of in vivo biotinylation of tagged proteins and streptavidin affinity capture to identify global targets of multiple factors involved in the transcriptional control of pluripotency in ESCs further highlights the power of proteomic approaches to define in a systematic fashion the protein-protein interaction and protein-DNA interaction networks operative in ESCs. In particular, affinity purification of biotin-tagged protein complexes coupled with LC-MS/MS (^{bio}SAIP-MS) and the ^{bio}ChIP-Chip method obviates reliance on low-affinity antibodies and allows for the generation of two independent data-rich resources with the same biotin-tagged cell lines and similar procedures, paving the way for highly efficient, large-scale proteomic studies in ESCs.

PROTEOMIC STUDIES OF ADULT STEM CELLS

Current Status of Adult Stem Cell Proteomics

Somatic stem cells have been identified within adult organisms, and are defined by their dual properties of self-renewal and differentiation. Unlike ESCs, however, adult somatic stem cells are restricted in their ability to give rise to cell types within a defined lineage. Over the last 20 years a large body of work has been compiled to further define these cells, develop rigorous isolation strategies, deduce their in vitroand in vivo functions, and establish transcriptional profiles. While these studies have greatly advanced the field, a complete understanding of the mechanisms that regulate self-renewal and potency within adult stem cells requires integration of multiple high-throughput platforms assessing transcriptomes, proteomes and protein interactomes.

Unlike ESCs, however, only a relatively small number of studies have ventured into proteomic profiling and protein interaction mapping of adult stem cells. The majority of studies in the field of adult stem cell proteomics have focused on three cell types: hematopoietic stem cells (HSCs), neural stem cells (NSCs) and mesenchymal stem cells (MSCs). There are many inherent challenges in pursuing proteomic studies using adult stem cells. With the exception of NSCs, which can be significantly expanded in vitro without loss of stem cell properties, most adult stem cell types cannot be maintained or expanded in culture without inducing changes in their potency. Thus, there are limits to the numbers of available input cells and unlike the development of global nucleic acid amplification for transcriptional profiling, currently there is lack of an effective protein amplification method.

Many of the initial proteomic efforts from in vitro expanded adult stem cells have utilized 2-dimensional gel electrophoresis (2-DE) as a front-end fractionation method prior to mass spectrometry (MS) analysis. There are several limitations inherent to this approach, including limited resolving power, poor representation of very large or small, basic or hydrophobic proteins, the requirement for relatively large amounts of sample, and statistical issues (different analysis algorithms generate divergent results). Combined data sets from these proteomic profiling studies reveal that the largest conserved group of proteins in adult stem cells are involved in energy metabolism. However, these data sets are largely biased by the methodology used and consequently may simply represent the most abundant proteins broadly expressed among these cell types.

Subsequent to these initial studies, several groups have taken advantage of the development of more sophisticated and unbiased proteomic techniques to gain new insights into adult stem cell biology. Development of sensitive iTRAQ methodology combined with MS analysis has allowed comparison of purified populations of hematopoietic stem and progenitor cells with as few as 1×10^6 input cells. Interestingly, results of this study suggest that HSCs, unlike their more differentiated progenitor counterparts, are adapted for anaerobic environments. These differences were not seen when the transcriptomes of these same populations were compared, strongly indicating that transcriptional profiling alone would not have been sufficient to deduce this novel aspect of HSC biology. Additionally, iTRAQ has been effective in defining a poorly characterized population of hematopoietic progenitor cells (Lineage⁻ c-Kit⁺ Sca-1⁻) as principally erythroid in nature. In the MSC field, 2-dimensional liquid chromatography (2D LC) or LC/LC fractionation followed by tandem MS/MS has been utilized to demonstrate that osteogenic differentiation of stem cells results from the focusing of gene expression in functional clusters rather than simply from the induced expression of new genes.

It has also been demonstrated that PTMs can significantly influence adult stem cell fate decisions. A quantitative phosphoproteomics approach, facilitated by SILAC technology, has been used to study the influence of growth factor signaling on MSC differentiation. Specifically, the mechanism by which two related growth factors (EGF and PDGF) differentially impacted MSC differentiation was found to be mediated by tyrosine phosphorylation (Kratchmarova et al., 2005).

As highlighted earlier in this review, the recent characterization of a functional protein interactome and transcription regulatory network in mESCs has yielded important new concepts in processes regulating development and stem cell pluripotency. While this type of intricate network has not yet been identified within adult stem cells, initial efforts towards this goal have utilized a proteomics approach to identify critical protein-protein interactions regulating self-renewal and differentiation. Using antibody-based purification of protein complexes, an elegant study by Lessard et al.

Has characterized an essential change in subunit composition of a SWI/SNF-like chromatin remodeling complex during differentiation of NSCs to post-mitotic neurons. Neural stem and progenitor cells express subunit proteins and BAF53a as part of the SWI/SNF chromatin remodeling complex, which are replaced by BAF45c and BAF53b as progenitors exit the cell cycle. Importantly, the essential nature of this subunit change for neural differentiation was functionally validated. Taken together, the proteomic profiling and protein interactome studies of adult stem cells achieved thus far highlight the fact that these methodologies can and will lead to novel insights into the underlying cell biology that would not be discovered using other means.

Future Directions of Adult Stem Cell Proteomics

The field of adult stem cell proteomics has a promising future. As new, improved and more sensitive methodologies become available, the limited numbers of obtainable adult stem cells will become less of a barrier. One very promising approach for a wide variety of applications in adult stem cell proteomic studies is use of protein microarrays or chips. These have the potential to identify protein-protein interactions, protein-phospholipid interactions, small molecule targets, and substrates of protein kinases, all while requiring a relatively small amount of starting material. The type of functional protein microarrays that have been used previously in yeast to study protein-protein interactions, specifically demonstrated to identify calmodulin binding proteins, stand to be particularly valuable in characterizing the adult stem cell protein interactome.

One of the major issues facing adult stem cell proteomics, i.e., cell heterogeneity, ironically stands to be greatly aided by proteomic work itself. It has been demonstrated that isolation of what is considered to be an enriched hematopoietic stem/ progenitor population (human umbilical cord blood $CD34^+$ cells) still results in significant proteomic heterogeneity between samples. Adult stem cell populations expanded in vitro are also not immune to this issue, as it has been shown that human bone marrow MSC lines have divergent self-renewal and lineage differentiation capacities. The way in which proteomics will be able to address these issues is through further characterization of cell surface antigens expressed specifically on various adult stem cell types, which will allow even greater prospective isolation capability and thus more homogeneous cell populations. This has been recognized in the HSC field, where transcriptome profiling enabled identification of the SLAM family of cell surface markers, which have improved means of HSC isolation. If transcriptome data is able to achieve moderate success to this end, there is vast potential to identify novel biomarkers through membrane proteomics. In addition, use of lineage-specific fluorescent reporters will allow isolation of more homogeneous cell populations. This strategy has been successfully employed in proteomic studies examining differentiation of mESCs to mesodermal/hemangioblast lineages with subsequent profiling using iTRAQ.

CONCLUDING REMARKS

The proteomics studies of embryonic, as well as adult, stem cells will complement characterization of these cells at the transcriptional level (transcriptome) and connect gene transcription and cellular phenotypes. The true challenge now is to integrate proteomics into the full spectrum of biological and biomedical research. Over the next decade, characterizing the proteome and interactome of stem cells through the identification of protein constituents, quantitation of protein concentration, dissection of protein interaction networks, and deciphering of transcriptional circuitry will provide a wealth of valuable information.

These data will enable an integrated systems-level analysis and modeling of the mechanisms regulating stem cell self-renewal and potency. Combined advances in stem cell biology and MS hold great promise for dissecting components or pathways that either stimulate proliferation and self-renewal or induce differentiation towards specific cells or tissues. Ultimately, this will provide a framework for understanding the underlying biology of stem cells, and allow precise manipulation and realization of the full clinical therapeutic benefits of these unique cells.

GENOME-WIDE TRANSCRIPTION FACTOR LOCALIZATION AND FUNCTION IN STEM CELLS

Stem cells are biologically and clinically important cells characterized by their ability to self-renew or differentiate into many cell types. The decision between self-renewal and differentiation is governed by extracellular signals, coupled to intracellular signaling cascades which activate transcription programmes. Hence, transcription regulation plays a determining role in conferring cellular identity and function. As intrinsic determinants, transcription factors provide an entry point for uncovering how stem cells attain their phenotype, and how lineage-specific differentiation is initiated. The step-wise maturation of stem cells into terminally differentiated cell types requires the timely activation of a cascade of transcription programmes governed by lineage-specifying transcription factors. The identification of key regulators in the hierarchy and their targets can provide clues into the manipulation of stem cells. This review examines the role of key transcription factors in various stem cell types, and emphasizes on their control of downstream gene expression, which ultimately contributes to cellular function.

INTRODUCTION

Regulation of Gene Expression

Transcription regulation constitutes a major component of biological activities in the cell. A key function of transcriptional activity is the control of gene expression necessary for the maintenance of a cellular state. In metazoan, all cell types, with the exception of mature germ cells, contain the same genetic information within the species. Yet, each cell type expresses a unique subset of all genes encoded. Differential gene expression can be controlled at many levels. One of the foremost steps occurs at the first stage of expression: transcription. Coding and non-coding genes contain promoter and enhancer sequences that are bound by transcription activators and repressors - transcription factors. A large proportion of TFs are DNA-binding proteins that localize to cis-regulatory DNA elements located upstream or downstream of their target genes. The levels and activities of TFs determine to a large part the extent of transcription. Hence, the active quantity of these regulators in turn must be tightly regulated. The regulatory mechanisms of TFs are diversified by the combinatorial action of upstream regulatory genes, post-translational modifications, ligand-dependant activation for nuclear hormone receptors, and protein-protein interactions. Two classes of TFs can be distinguished in the eukaryotes. General transcription factors (GTFs) bind to a core promoter usually close to the transcription initiation site, together with the assembly of the pre-initiation complex, while sequence-specific TFs bind regulatory DNA elements at varying distances away from the transcription start site. The occupancy of transcription activators on a promoter recruits the transcription initiation machinery which consists of RNA polymerases and more than 50 other components.

This is followed by promoter clearance, elongation, and the termination of transcription. Other components of the post-transcription machinery are also involved in augmenting the overall levels of gene transcripts in the cell. These include regulated mRNA decay which is achieved through the interactions between trans-acting factors with the mRNA's structural components, the sequence dependent mRNA stability, and the association of mRNA with general or mRNA-specific RNA-binding proteins. Therefore, transcription regulation, coupled with post-transcriptional modifications, help determine the base-line of differential gene expression profile that confers cellular identity.

A Stem Cell Window into Transcription Factor Function

Stem cells are well-suited for the study of transcription regulation and TF function. As these remarkable cells are characterized by their ability to self-renew as well as differentiate into functional cell types, they respond to external stimuli and intracellular perturbations that are observable and detectable. During the course of animal development, cells transit from a pluripotent state to one of many committed lineages. Several developmental stage-specific stem or progenitor cells have been detected and isolated from the mammalian embryo, as well as from post-natal and adult tissues. These include lineage-committed multipotent stem or progenitor cells such as hematopoietic stem cells (HSCs), neural progenitor cells (NPCs), epithelial stem cells and mesenchymal stem cells. Of particular interest are cells of the embryonic inner cell mass (ICM) and their derivatives, the embryonic stem cells (ESCs), which are capable of giving rise to all subsequent cell types of the fetus.

This makes ESCs highly promising in regenerative medicine. An important requirement for directing differentiation of stem cells is to first understand how the undifferentiated 'stemness' state is established and maintained. This would give us clues on how to isolate and expand stem cell population, as well as control their cell fate specification either chemically or genetically.

At the heart of cellular differentiation programmes lies a plethora of highly regulated TFs that enable the alteration of cell states in response to extracellular signaling cues. For many decades, the study of TFs as regulators of downstream effectors has been limited in scope and comprehensiveness. Deciphering the sequential order in which top-tier regulators are activated and identifying what are the extensive repertoire of target genes has been difficult.

Different approaches have been employed to address these issues. During cell fate specification, the activation of a hierarchy of TFs has been inferred from the analyses of mutant animals. This is best exemplified in gene knockouts of TFs involved in specific tissue development such as myogenesis, hematopoiesis and neurogenesis.

Downstream perturbations arising from the loss of a key TF regulator have provided clues for the identity of regulated target genes. In recent years, the development of microarray technology and sequencing of many vertebrate genomes has facilitated in capturing gene expression profiles on a global scale and the unraveling of transcription regulation hierarchy. One strategy involves perturbing the TF of interest by either over-expression or depletion, followed by analysis of the resultant changes in gene expression profiles. However, whether differentially expressed genes are directly bound and regulated by the TF, or indirectly affected through the perturbation of intermediate genes, cannot be distinguished.

Therefore, alternative approaches are required to map TF binding sites in the genome accurately and efficiently.

Interrogation of Transcription Factor Binding Sites

Chromatin immunoprecipitation (ChIP) allows for the identification of physical interactions between proteins and DNA in the context of chromatin. An overview of the experimental procedure is depicted in Figure. ChIP involves chemical cross-linking of DNA-protein interactions in living cells to capture TFs to their cognate binding sites. Cross-linked chromatin is fragmented and protein-DNA complexes can be enriched by immunoaffinity pull-down of the desired protein. DNA-protein cross-links are then reversed, and the enriched DNA can be purified for analyses.

When combined with high-throughput detection instrumentations, ChIP can provide a comprehensive view of protein-DNA interactions. Three major strategies have been employed for locating TF occupancy in a genome-wide manner. Briefly, ChIP-chip involves the amplification of ChIP-DNA followed by hybridization to whole-genome promoters or customised DNA microarrays. The difference in fluorescent signal intensity of ChIP-DNA over control DNA reflects the relative enrichment of the TF on genomic binding sites. ChIP-PET (pairedends di-tags) represents a sequencing-based method for detecting enrichment of ChIP-DNA.

It involves the construction of a concatenated DNA fragment plasmid library which is then sequenced. Mapping both ends of a fragment has the advantage of accurately decoding the identity of a gene fragment, and this method is not constrained by array-based limitations such as probe performance and genome coverage. Recently, the application of ChIP-seq (sequencing) demonstrated considerable advantages over the other methods. ChIP-seq involves direct sequencing of ChIP-DNA which has been size-selected and amplified. The method is simple, requires small amounts of ChIP-DNA, has deeper coverage, and is more efficient. Therefore, the availability of these methods has greatly enhanced our understanding of the complexity that governs transcription regulatory networks in higher eukaryotes. In addition, genome-wide ChIP analysis has also been useful for the study of non-TFs that associate with, or are components of, the chromatin. These proteins include modified histones and epigenetic regulators such as Polycomb complexes.

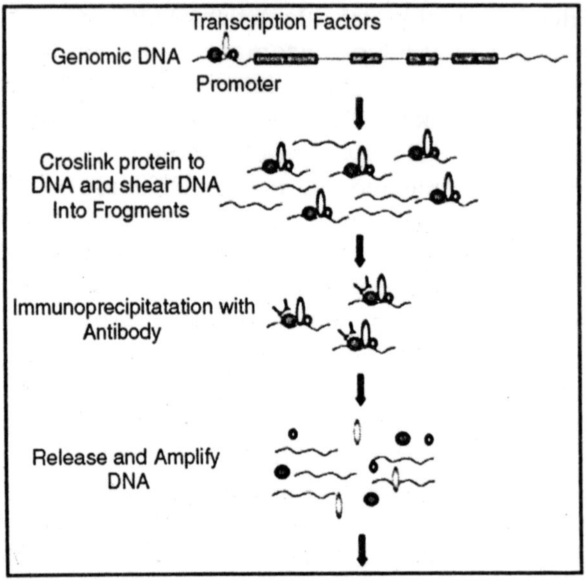

Fig. Genome-wide Transcription Factor Location Analysis.

Transcription factors binds to the cis-regulatory sequences of chromatin. To identify the genomic loci bound, the protein is cross-linked to DNA, and the chromatin is sheared. Chromatin immunoprecipitation is used to purify protein-DNA complexes using an antibody of interest. The DNA is released and analysed. The occupancy of the TF across the genome can be accessed by three high-throughput methods: ChIP-chip, ChIP-PET and ChIP-seq.

Genome-wide TF localization analysis has provided invaluable insights into the functions of regulators involved in cell homeostasis, cell fate transition, immune defence and cancer We and others have employed this strategy to examine the transcription circuitry of ESCs which is governed by a hierarchy of core factors. In uncovering and reconstructing the key features of this circuitry, we further dissected the significance and contributions of these transcriptional components to ESC function. The spatial and temporal specifications of stem cells in the embryo and adult somatic tissues have raised important questions pertaining to their establishment and maintenance. In particular, the contrasting role of TFs either as regulators in stem cell maintenance or directing differentiation is of immense interest.

Thus, mapping the stem cell transcription network can provide useful clues into their manipulation. Here, we discuss key findings into the deconstruction of molecular circuitries governing stem cell function, with an emphasis on ESCs. Knowledge drawn for these studies has further led to the synthesis of key concepts in helping us understand the complexity of gene regulatory network in stem cells.

KEY REGULATORS OF PLURIPOTENCY

During development, the formation of the ICM in the mammalian blastocyst is a major landmark that represents the specification of pluripotent cells. ESCs have been derived from human and mouse blastocysts, and are similar in many developmental aspects. Although differences exist in their culture and maintenance, many molecular determinants governing their undifferentiated states are shared. Homeodomain TFs are evolutionary conserved and participate in cell fate determination in many organisms. It is exclusively expressed in blastomeres, pre-implantation embryos, primodial germ cells, the ICM and a narrow range of cultured cell As a TF, Oct4 can activate or repress gene expression and interact with other molecules such as the HMG-box TF, Sox2, in pluripotent cells.

Genetic studies have confirmed the importance of Oct4 during embryonic development and in maintaining pluripotent ESCs. Nanog, another homeodomain-containing TF expressed abundantly in ESCs, primodial germ cells and embryonal carcinoma cell lines, is also necessary for ESC function. In the mouse embryo, Nanog first appears in the compacted morula, becomes localized to the ICM, and is finally restricted to the epiblast. In Nanog$^{-/-}$ embryos, there is a lack of epiblast and primitive ectoderm formations, indicative of the absolute requirement for Nanog; and Nanog$^{-/-}$ ESCs manifest a reduction of pluripotency and re-specification to the extra-embryonic endoderm lineages.

Strikingly, a recent observation has clarified the role of Nanog further. While Nanog is necessary for inner cell mass formation and primodial germ cell specification, it appears dispensable in ESCs since Nanog$^{-/-}$ ESCs persist in culture although these cells are more prone to differentiation.

A host of other TFs also appear crucial for ESC function. The neuronal repressor REST is highly abundant in mouse ESCs and its depletion either through RNA interference (RNAi) or heterozygous deletion leads to a loss of self-renewal. REST is a major transcriptional repressor of neurogenesis, and the activation of REST-regulated genes can convert neural progenitor cells to neuronal phenotypes. But, the manner in which it regulates gene expression in ESCs is unclear.

Another regulator, Sall4, is an important activator of Oct4, and its loss leads to cell fate re-specification into trophoblast stem cells. This effect is recapitulated during pre-implantation development where down-regulation of Sall4 causes mis-expression of the trophoblast stem cell determinant, Cdx2, in the ICM. The use of integrated approaches combining transcriptome profiling, genome-wide ChIP analyses, as well as large-scale gain- or loss-of-function has further led to the identification of an increasing number of developmentally important TFs. In these studies, Tcl1, Tbx3 and Esrrb have been identified as novel regulators in ESCs. However, the precise manner in which these genes exert their regulatory effects has not yet been elucidated.

THE TRANSCRIPTIONAL LANDSCAPE IN ESCS

Initial studies into transcriptional regulation by Oct4, Sox2, Nanog and Sall4 have revealed a small number of genes bound by these factors. These limited observations are simplistic, and do not provide a comprehensive understanding into the complex framework of transcription hierarchy where regulators tend to exert far-reaching effects on a vast array of genes, as observed from yeast studies. In animals as diverse as C. elegans and Drosophila to Xenopus, mouse and man, activating and inhibiting signals establish a dynamic cascade of coordinated gene expression in response to extrinsic signals and intrinsic programming.

The integration of these instructions occurs in the nucleus through combinations of signal-activated and tissue-restricted TFs binding to, and controlling, cis-regulatory modules of genes. With the complexities arising from such intricate interactions, the ESC model, which shows a remarkable sensitivity to genetic perturbations, thus represents a suitable system for probing TF function and regulatory dynamics.

Auto-regulation and Feedback Loops Maintain Gene Expression

One emergent theme arising from transcription regulation in ESCs is the revelation that a core set of factors act together in a tightly regulated framework to control similar as well as divergent downstream genes. Auto-regulation by top-tier TFs is a key feature. The pluripotent molecules Oct4,Sox2, Nanog and Sall4 auto-regulate through binding to their own promoters. On its own, each factor maintains its expression that is individually required for ESC function. Collectively, all four factors regulate one another through the interconnectivity of positive feedback loops. The generation of these feedback circuits reinforces the transcription activity of each factor, and maintains appropriate levels of the necessary factors.

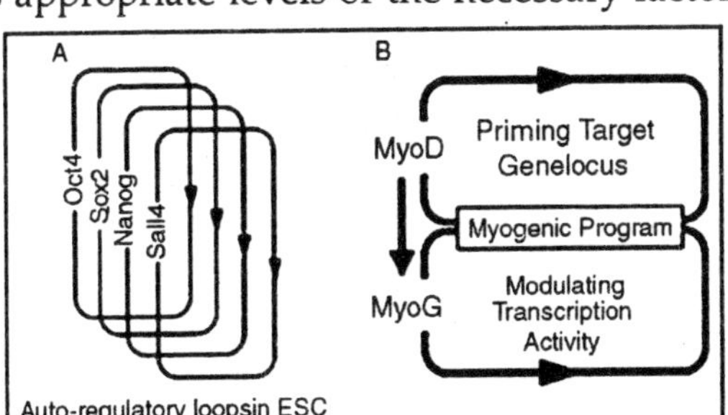

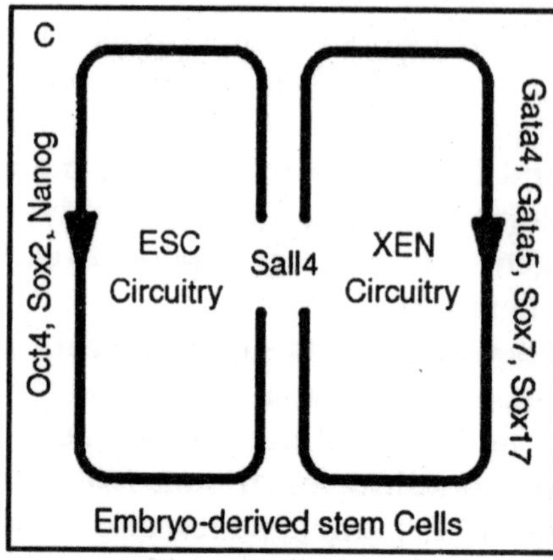

Fig. Transcription Programmes in Stem Cells.

Transcription factors control the molecular circuits that govern stem cell identity. A. In ESCs, key pluripotency-associated factors Oct4, Sox2, Nanog and Sall4 form auto-regulatory loops that maintain their own transcriptional activity and gene expression. The stability of this set of core factors is further enhanced through the interconnectivity of these loops.

B. During myogenesis, MyoD is an upstream regulator which activates the expression of MyoG. Both TFs are necessary for the myogenic programme, and regulate a common subset of genes in a temporal manner. MyoDmust initiate chromatin remodeling of genes for these to be subsequently bound and regulated by MyoG. C.

The same TF, Sall4, have dual roles in controlling discrete circuits in the different embryo-derived stem cell lines, ESCs and XEN cells.

In the respective cell types, Sall4 regulates distinctive sets of key factors that contribute to the specific cellular identity. In ESCs, Sall4 activates Oct4, Sox2 and Nanog; all of which positively feedback onto the Sall4 promoter.

In XEN cells, Sall4 activates Gata4, Gata6, Sox7 and Sox17 which are necessary for the undifferentiated state. Whether these regulated factors directly feedback onto the Sall4 promoter is not known.

Models of Transcription Regulatory Network

Key Concepts:Auto-regulatory and feedback loop

Remarks:

- Transcription factors (X and Y) auto-regulate by binding to their own promoters.
- X and Y also bind the promoters of each other in a feedback loop.
- Auto-regulation and feedback loops are features of 'master' regulators in the transcription network of eukaryotes.
- Provides enhanced stability of gene expression that maintains high levels of both factors.
- *Examples*: Oct4, Sox2, Nanog and Sall4in ESCs; HNF1a, HNF4a and HNF6 in hepatocytes.

Model:

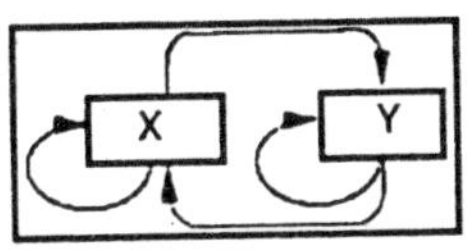

Key Concepts: Feed-forward loop

Remarks:

- Transcription factor (X) controls another regulator (Y), and both of them regulate downstream target gene (Z).
- Multiple positive inputs provide consistent activity in regulating a common target, rendering it insensitive to transient changes in individual input.
- Particularly important in stem cells to maintain either a stable self-renewing or differentiated state.
- *Examples*: In ESCs, Nanog regulatesOct4 and Sall4, where Oct4 also controls Sall4; upon differentiation, Sall4 is down-regulated partly owing to the loss of Nanog and Oct4.

Model:

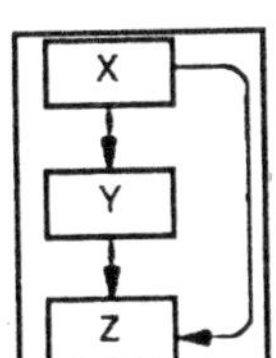

Key Concepts: One transcription factor → Multiple target genes

Remarks:

- A regulator (X) controls the expression of many target genes.
- Dominant key factors control a wide repertoire of genes in a positive or negative manner, which may require the activity of co-factors.
- The sum of activity of all downstream targets represents the cellular state.
- *Example*: REST represses extensive neuronal lineage-associated targets in non-neuronal cells; HNF1α controls hepatocyte biochemistry-related genes in liver; Oct4 activates ESC-associated genes and represses differentiation-associated genes in ESCs.

Model:

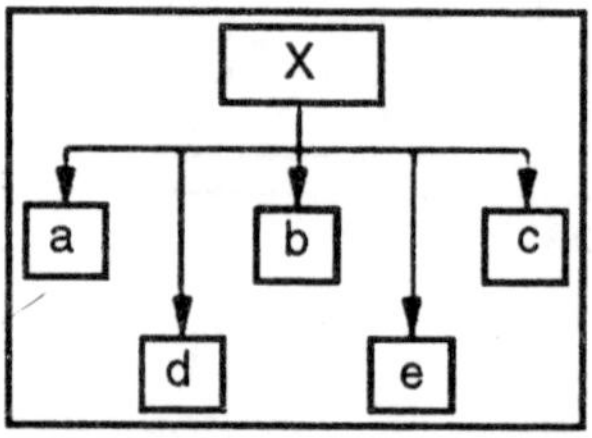

Key Concepts: Core set of transcription factors → Common target genes

Remarks:

- Transcription factors (X, Y and Z) forms a central regulatory hub.
- All three factors tend to bind common target genes, in addition to other targets.
- Core factors are known to regulate a set of common downstream regulators which themselves are crucial for the maintenance of a particular cellular state.
- *Examples*: Oct4, Sox2, Nanog regulates common genes in undifferentiated ESCs; MyoD and MyoG maintains the expression of common genes during myogenesis in a temporal manner.

Model:

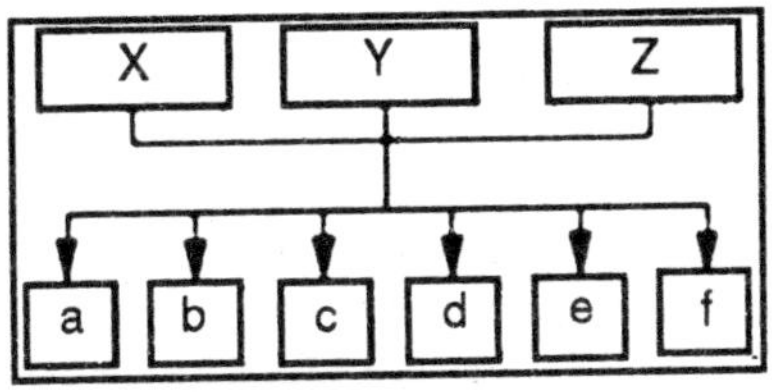

Key Concepts: Combinatorial inputs → Transcription output

Remarks:

- Transcription factors comprising of activator (A) and repressor (R) bind to a common target gene.
  The combinatorial input of activators and repressors determine the transcription output of Y.
- Output may be binary (On/Off) or graded (High/ Low).
- Repressors may behave as 'transistors' which moderate gene expression levels.
- *Example*: Regulation of the Oct4 gene activity by the activator Nanog and the repressor Tcf3 in ESCs.

Model:

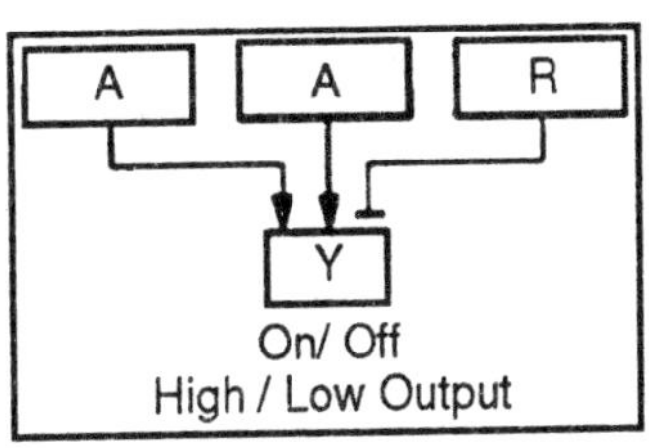

Key Concepts: One transcription factor → Distinct circuitries

Remarks:

- The transcription factor (X) may control different and/ or overlapping sets of target genes in a cell-type dependent manner.
- The regulation of dissimilar gene sets is responsible for a varied role in different cell types or cellular contexts.
- The control of a similar set of genes may suggest a common function for the factor in different cell types.

- *Example*: Sall4 maintains lineage-distinct ESCs and XEN cells by binding to different sets of genes; c-My regulates similar proliferation-associated genes in ESCs and cancer cells.

Model:

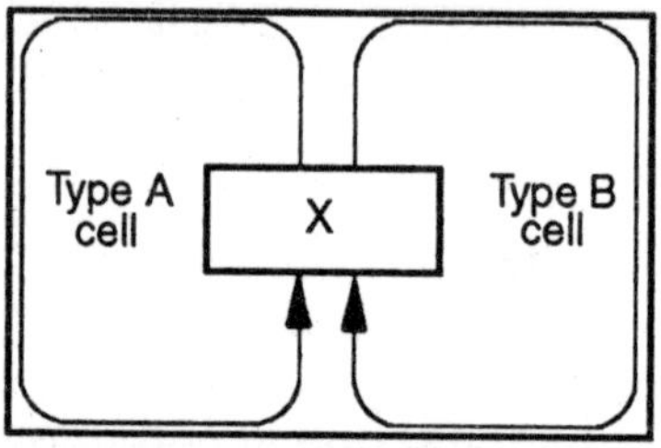

Key Concepts: Co-factor dependent recruitment to targets

Remarks:

- The activity and target gene occupancy profile of a transcription factor (X) is dependent on its association with specific co-factors (α or β).
- Co-factors can mediate recruitment of X to specific target genes that may be motif (DNA sequence) dependent.
- *Example*: FoxA1 recruitment to differential targets in breast and prostate cancer cells is mediated by the ERα and AR, respectively.

Model:

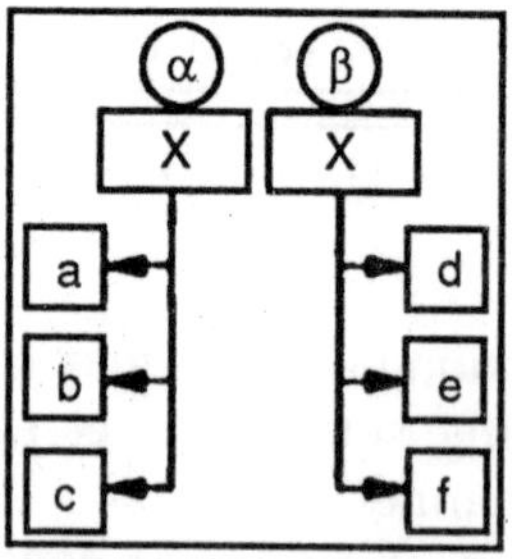

Key Concepts: Condition-dependent recruitment to targets

Remarks:

- The transcription factor (X) regulates a set of target genes in the native state of the cell.
- Upon stimuli, the binding dynamics of X may

become expanded or altered to include either a larger repertoire of targets relevant for a specific response, or an entirely different set of genes.

- The expression of co-factors induced by stimulation may be necessary for recruitment to targets.
- *Example*: Stimulation of monocytes by LPS expands the binding profile of $NF_{-k}B$ which is partly dependent uponE2F1.

Model:

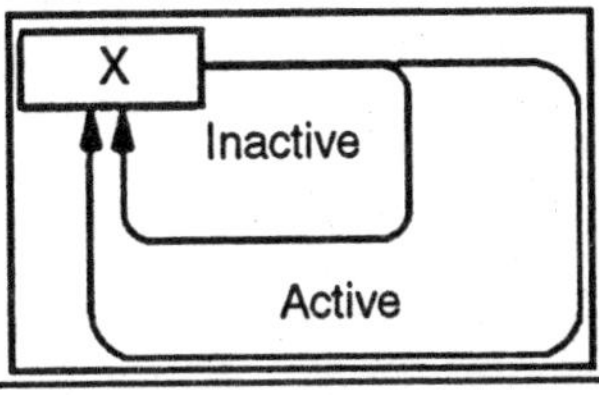

Recently, the transcription landscape has been extended through the analyses of a greater number of factors associated with ESC pluripotency, as well as those implicated in the reprogramming of somatic cells to induced-pluripotent stem (iPS) cells.

The extended repertoire examined reinforces previous observations that auto-regulatory and feedback loops are prevalent features of ESC transcriptional landscape. In one study, the interconnectivity of nine TFs has been analysed. Of these regulators, the promoters of Oct4, Sox2, Nanog and Dax1are bound by at least four factors including itself, suggestive of a robust positive feedback network. Strikingly, new regulatory foci that include Sall4, Rif1, REST and Dax1, bound by multiple TFs, have also been uncovered, integrating these hubs to the core circuitry.

Auto-regulation and feedback loops appear to confer several advantages. A core set of factors that maintains its own expression, as well as that of others in the vicinity, enhances the stability of gene expression.

This is especially vital to ESCs since the prolonged loss of any key factor is sufficient to cause irreversible differentiation. Based on empirical evidence that have emerged from biological studies, the computational notion of a 'bistable' switch arising from the positive feedback loops comprised of Oct4, Sox2 and Nanog has been proposed. It is thought that such a mechanism stabilizes the expression levels of all three factors, which through their regulatory control of downstream targets, appears extremely tolerant to parametre change and could counter input noise.

Feed-forward Regulation of Common Target Genes

The second observation is the regulation of common target genes by ESC regulators. Oct4, Sox2 andNanog form feed-forward loops that involve over 350 protein-coding genes in human ESCs.

In the mouse counterpart, Oct4 and Nanog occupy approximately 1,000 and 3,000 genomic binding sites respectively. Many of these Oct4-bound loci also contain Sox2sites, confirming both factors act in tandem. Significantly, Oct4 and Nanog positively control downstream genes important for maintaining pluripotency and inhibiting differentiation.

These included Foxd3, Setdb1,Rif1, Esrrb and n-Myc which help maintain pluripotency. A cascade of other pathways that are connected to self-renewal, genome surveillance and cell fate determination are also implicated. In addition to their role as activators, Oct4 and Nanogcan repress differentiation-associated genes. One key example repressed by Oct4 is Cdx2, necessary for trophectoderm formation.

Additionally, integrated analysis of a larger set of TFs revealed that target genes co-occupied simultaneously by multiple other factors are found to be generally active in ESCs, and repressed upon differentiation; whereas those bound by a single factor alone tend to be inactive or repressed.

This suggests multiple inputs may be necessary for a concerted activation signal, possibly through the assembly of activation complexes.

For single inputs, these could be associated with co-repression complexes, as revealed by protein network analysis where there exist multiple connections between core factors and repressive chromatin complexes such as NuRD and Polycomb. Several TFs may bind to common gene promoters, or at the same genomic loci. A. In ESCs, different combinations of pluripotency-associated factors, Oct4, Sox2, Nanog, Klf4, Nacl and Dax1, occupy the promoters of common genes. In this multiple co-occupancy where at least five factors are bound, it typically leads to transcription activation of the target genes, highlighting a synergistic effect that is mediated by a combined positive transcriptional input.

B. Multiple TFs are also found to be located on enhancers where these dense 'hotspots' are referred as enhanceosomes. As many as 12 factors are found to occupy the enhancer of Oct4, which is a key ESC regulator.

Several of these TFs have been shown to biochemically co-purify as part of a larger complex, supporting the synergistic activation model.

Interestingly, discrete feed-forward loops comprised of divergent target sets bound by different subsets of TFs can be discerned.

In a separate study involving the genome-wide mapping of 13 TFs (Oct4, Sox2, Nanog, Klf4, c-Myc, n-Myc, Esrrb, Zfx, Tcfcp2l1, Stat3, Smad1, E2f1 and CTCF), it is striking that whereas targets of Oct4, Sox2, Nanog, Stat3 and Smad1 appear more closely associated based on hierarchical clustering, targets of c-Myc, n-Myc and Zfx fall into a separate grouping.The interrogation of chromatin occupancy for a large set of TFs, not limited to promoters, has surprisingly revealed genomic regions, or hotspots, which are co-occupied.

If only a few factors were examined, the existence of these hotspots would not have been evident. These nucleoprotein complexes comprised of a combination of TFs binding to enhancer DNA are referred to as enhanceosomes. It is furthermore striking that the densest binding loci occupied by 12 of 13 factors is the previously characterized distal enhancer of Oct4. Several factors in this complex have already been reported to interact with one another. Given the limited understanding of enhancer functions, in-depth analyses of enhanceosomes, or common target loci, could provide valuable insights into key regulatory regions unique to ESCs. The reliance on feed-forward loops has several advantages. When more than one positive regulator control a common gene, either directly or indirectly through intermediates, the multiple inputs provide consistent activity that render it relatively insensitive to transient changes in individual input strength.

When these regulators have either a positive or negative function, the feed-forward loop then behaves as a sensor that can respond rapidly to the balance of the inputs. In that sense, the upstream genetic cis-regulatory elements can be viewed as a 'transistor' for receiving and moderating the input signals of the TFs which may carry either an activating or repressive function.

This concept of feed-forward loops appears particularly vital to stem cells which must react appropriately to either self-renewal or differentiation cues.

Much less is understood about the consequences of having elevated levels of pluripotency factors. As exemplified by Oct4 and Nanog, their elevation at non-physiological levels dramatically alters ESCs.

An up-regulation of Oct4 causes ESCs differentiation towards the mesendoderm lineage while an increase in Nanog limits their ability to differentiate. Furthermore, the ectopic mis-expression of Oct4 in somatic tissues of adult mice can lead to dysplastic growth in epithelial tissues and block progenitor cell differentiation. Thus, it is imperative to understand the fine-tuning of appropriate transcription output by repressors.

Fine-tuning the Transcription Output

The negative regulation of pluripotency-associated genes is poorly understood and largely under-appreciated. The tumor suppressor p53, for example, suppresses transcription of Nanog and induces differentiation in response to DNA damage, suggesting its role in maintaining genetic stability and, the orphan nuclear receptor GCNF, was reported to repress Oct4 and Nanog during retinoic acid-induced differentiation of ESCs. Recent studies have further highlighted the role of the Wnt TF, Tcf3, as a potent repressor of Oct4 and Nanog. The Wnt signaling cascade is frequently implicated in the establishment of stem cell identity, as well as directing cell fate commitment.

But, the precise mechanism in which its downstream effectors, Tcf/Lef factors, mediate these decisions is not known. Interestingly, the depletion of Tcf3 limits the ability of ESCs to differentiate.

Part of this is attributed to the relief of repression on Oct4 and Nanog, whereby their levels become induced. Comprehensive analyses of Tcf3 binding sites across the genome surprisingly revealed that this repressive factor share many common targets with Oct4, Sox2 and Nanog.

The close proximity of these factors on co-occupied genes further suggests an integrative regulatory function, and many of their common targets have roles in ESC self-renewal.

It appears conceivable that Tcf3 counters the effects of positive regulators on these genes as Tcf3 depletion induces the up-regulation of these targets. The conglomeration of data indicates that fine-tuning the dosage of ESC factors is a crucial component in transcription regulation.

Currently, the interaction between activators and repressors on the same gene promoter is not well-understood, much less their complex dynamics on a global scale. The delicate balance between positive and negative inputs necessary to dictate the transcriptional output needs to be better established before quantitative evaluation and predictions of the transcription circuitry are possible.

The recruitment of co-repressors, such as Grouchos/TLEs or activators such as β-catenin in the case of Tcf/Lef factors confounds our understanding on how TFs act as molecular switches in a combinatorial manner.

In-depth studies of interactions among transcription regulators and the global mapping of an increasing number of ESC factors are pertinent in the systems approach to dissecting the stem cell identity, as exemplified in a computational study model.

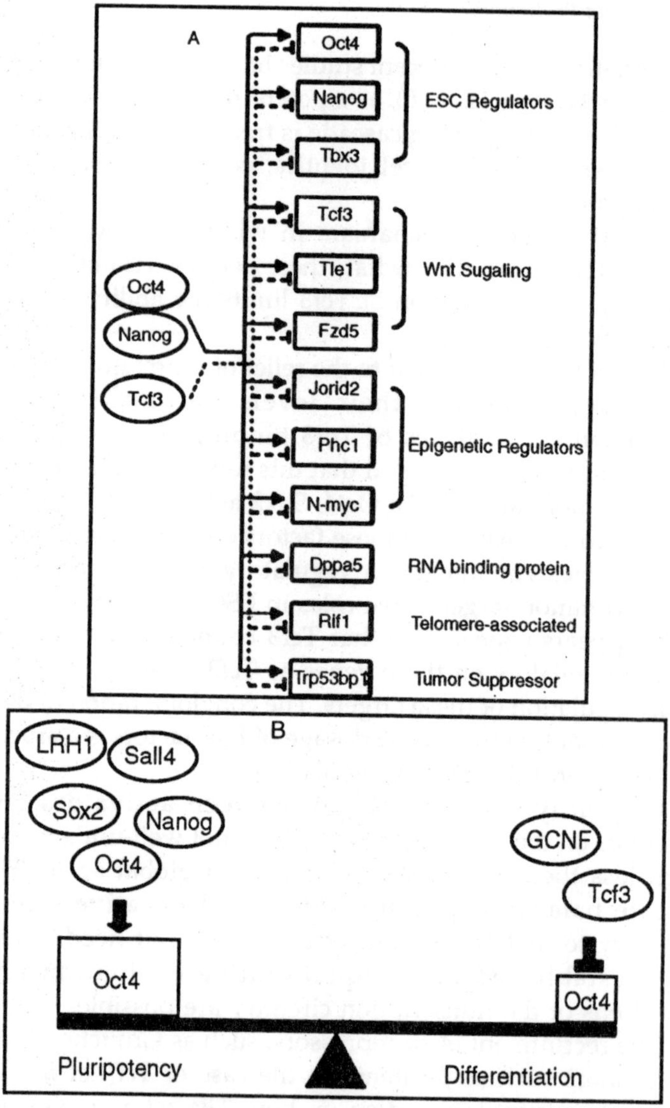

Repressors Modulate Transcription Output

A. Oct4, Nanog and Tcf3 co-occupy a common set of pluripotency-associated factors in ESCs. Whereas Oct4 andNanog may activate genes required for pluripotency, Tcf3 behaves as a repressor that also binds these common genes, thereby controlling the overall gene expression levels. B. At the single gene level, the transcription output is partly determined by the combination of factors that activate or repress transcription.

The well-characterized Oct4gene cis-regulatory regulatory region is bound by several TFs that act as either activator or repressor. In the undifferentiated state, Oct4, Sox2, Nanog, Sall4 and LRH1 activate the promoter while Tcf3 and GCNF repress it. The balance of molecules acting on this critical gene controls the cell fate of either self-renewal or differentiation.

Transcription factor pleiotrophy

Surprisingly, the role of Tcf3 extends beyond its control of pluripotency-associated genes. It also binds the promoters of homeodomain genes which include members of the Fox, Tbx and Sox gene families known to regulate early developmental programmes.

Tcf3 appears necessary to prevent the premature activation of differentiation programmes. Whereas polycomb group (PcG) proteins act as general transcription repressors that blocks the expression of a large cohort of developmental regulators,Tcf3 binds to two categorically discrete sets of regulators in the same stage-specific context, i.e. undifferentiated ESCs.

Although both sets of targets become de-repressed upon the loss of Tcf3, it remains unclear why Tcf3-deficient cells tend to self-renew rather than differentiate. A plausible explanation may be attributed to the functional dominance of ESC stemness-associated over differentiation-associated genes, as observed during somatic cell reprogramming by fusion with ESCs.

This pleiotropic control by a TF in a genome-wide manner is puzzling, and implies critical roles for repressors. Interestingly, it has also been demonstrated that Wnt signal activation can induce ESC-associated targets of Tcf3, possibly due to the replacement of co-repressors with the co-activator β-catenin which associates with Tcf3. It, however, remains unresolved whether there is indeed a direct protein-protein interaction between Tcf3 and β-catenin upon Wnt stimulation, or that β-catenin may interact with other Tcf/Lef factors to activate common genes.

Functional validation of transcriptional targets

A major limitation of genome-wide studies is the generation of large interactome datasets that are not functionally tested in a biologically relevant manner. Technically, it is demanding to prove a causative relationship between gene occupancy and transcription activity one gene at a time.

Although an alternative method is to correlate global transcript changes with factor activity for implying a direct transcription control, many studies surprisingly revealed that a large proportion of these gene targets do not appear directly regulated. Transcription factor over-expression or depletion does not necessarily alter the transcript levels of target genes.

This, therefore, raises an important question of whether TF occupancy alone is sufficient to infer function in an overall network, or is the association with co-factors and protein complexes really the critical determinant. Furthermore, the epigenetic state of chromatin in altering permissiveness to transcription activity is another major consideration not discussed.

One approach which clarifies the role of transcription targets is through gain- or loss-of-function studies to test if putative candidates are physiologically relevant. The availability of extensive gene expression libraries has made this type of validation possible.

It is now feasible to establish high-throughput platforms for introducing a library of over-expression or RNAi plasmid constructs into stem cells, accompanied by a suitable assay method for assessing gene function.

For genetic studies in ESCs and gene function during animal development, the accessibility of gene-trap mouse ESC lines has made it easier to analyse gene functions in the animal. Hence, genomic discoveries create a framework which supports hypothesis-driven candidate gene function analysis.

TRANSCRIPTIONAL LANDSCAPE IN SOMATIC STEM CELLS

Distinct Transcription Factors Direct a Common Programme

The global transcription landscape in many somatic stem cells is poorly understood. Part of this limitation is the inherent difficulty of obtaining a large number of homogenous bona fide somatic stem cells in vitrofor analyses. Most somatic stem cells, unlike ESCs, arrest at the interphase of cell cycle and do not proliferate rapidly unless induced. Limited knowledge of the culture condition has prevented efficient expansion. Notwithstanding, unipotent and multipotent stem or progenitor cells of several tissues can be isolated and cultured. During myogenic differentiation, the convergence of lineage-defining TFs on a common set of targets amplifies the recurrent theme of a core transcription network.

Precursor myoblast cells proceed through irreversible cell cycle arrest, followed by an increase in muscle genes expression. This process is orchestrated through a series of transcription networks governed by distinct myogenic TFs. Several basic helix-loop-helix (bHLH) family members play discrete roles.

MyoD and Myf5 are necessary to specify the initial skeletal muscle lineage, where the ectopic expression of MyoD alone is sufficient to force non-muscle cells to complete the myogenic programme. The subsequent expression of MyoG causes the terminal differentiation of specified muscle cells and Mrf4 appears to act as a specification as well as differentiation factor (Kassar-Duchossoy et al., 2004). In the skeletal muscle programme, genome-wide identification of MyoD and MyoG binding sites revealed that both factors have distinct and sequential regulatory roles on a similar set of genes promoters. MyoD initiates chromatin remodeling but is insufficient for gene expression, and MyoG can only bind genes previously initiated by MyoD. The shared consensus DNA binding sequences of MyoD and MyoGfurther confirms their convergence.

This sustained temporal activation of similar targets by two stage-specific factors is surprising because previous evidence suggest that MyoD andMyoG act on separate sets of myogenic gene promoters thereby promoting a step-wise maturation. The convergence of both factors on common targets could not have been revealed in limited candidate gene analyses, as a comprehensive approach to mapping the transcription landscape during myogenesis is necessary to resolve these differences.

Spatial and Temporal Expression of Transcription Factors

The specification of the vertebrate central nervous system is defined by neural progenitor/stem cells (NPCs/NSCs) which proliferate, self-renew, and give rise to neurogenic and gliogenic lineages. A few pro-neural genes which encode the bHLH TFs are necessary and sufficient to initiate the formation of neural lineages, and also to promote the generation of progenitors that are committed to differentiation. Neurogenins (Ngn1/2/3) and NeuroD are key regulators of vertebrate neurogenesis.Ngn1 or Ngn2 mutant mice lack cranial and spinal sensory ganglia and ventral spinal cord neurons, while NeuroD mutants manifest the loss of cerebellar and hippocampal granule cells, inner ear sensory neurons, and retinal photoreceptor cells.

The expression of pro-neural genes in NPCs is transient as these must be down-regulated before progenitors exit the proliferative zone of the neural tube and differentiate. Hence, pro-neural genes appear to promote complete neuronal differentiation through the induction of downstream regulators. The direct transcriptional targets of pro-neural genes has not been systematically defined, but indirect evidence in Xenopus suggests that Ngn1/2 and NeuroD may regulate many genes such as Nhlh1, NeuroD4 and Hes, which are related to neurogenesis. In pancreatic islet cells, NeuroD also appears to be a downstream mediator of Ngn activities.

Unlike ESC regulators which must remain constitutively elevated, the control of NPC fate by pro-neural genes occurs in two phases. The first involves a reversible selection of progenitors which are not committed to differentiation when pro-neural genes are expressed at a low basal level. This is followed by a second phase whereby elevated levels of pro-neural genes force an irreversible commitment of progenitors to differentiate. Although signaling pathways that include Notch, sonic hedgehog and bone morphogenic protein (BMP) provide extrinsic cues for the initiation of pro-neural genes, it is likely that other positive feedback mechanisms mediated by downstream transcription targets help increase or maintain their appropriate expression in the selected progenitors.

Subsequently, upon the induction of differentiation, lineage-restricted genes become elevated and direct the progenitors to adopt their committed fate. Even though pro-neural and neuronal-differentiation genes perform distinct functions, the structural similarities between them have raised the question of whether their disparate functions may be due to the consequences of stage-specific and spatial-dependent expression in the neural lineages, or intrinsic biochemical differences such as protein activity and/or stability. In the spinal cord, the spatial expression of neural bHLH proteins Math1, Ngn1 and Mash1 in distinct dorso-ventral progenitor domains is responsible for controlling the specification of interneuron subtypes.

Observations from myogenic differentiation appear to support the view that temporal expression of structurally related TFs plays a major role in defining the transcription programme of either self-renewal or differentiation. Therefore, the expression specificity in a spatial and temporal context, rather than unique biochemical properties, of bHLH pro-neural genes may determine the control of differential target genes.

Several members of the Sox family of TFs also mark neural progenitor cells throughout the central nervous system, as well as maintaining self-renewing NPCs in vitro. The constitutive expression of Sox2inhibits neuronal differentiation and results in the maintenance of progenitor characteristics, whereas its loss results in NPCs exiting cell cycle and the onset of neuronal differentiation. The multipotent NSCs obtained from mouse embryonic neuroepithelial cells express Sox2, along with the neural cell marker Nestin. Strikingly, Sox2 is also a determinant of ESC function. This underlying importance of the same factor in two stem cell types could suggest that either the temporal and lineage-specific expression of a TF determines their cell fate, or Sox2 could be a 'stemness' gene common to certain stem cells. Nevertheless, the lack of information on Sox2 transcription targets in NSCs hampers a direct comparison of the mechanisms which it may utilize in different stem cells.

A Transcriptional link Between Hematopoiesis and Leukemia

During hematopoiesis, the establishment and maintenance of the entire blood system depends on rare hematopoietic stem cells (HSCs) residing in the bone marrow. The balance between self-renewal, and the differentiation of HSCs into lineage-restricted progenitors, and ultimately into diverse blood cell types, is intricately controlled through extracellular and intracellular signaling cues.

These feed into the activation of regulators, which establish complex transcription networks. Briefly, HSCs can give rise to two major lineage-restricted progenitor cell types: the common myeloid progenitor which diversifies into myeloid cells comprised of monocytes, neutrophils, eosinophils, erythroids, mast cells and megakaryocytes; and the common lymphoid progenitor which produces the B and T lymphocytes. Several TFs maintain HSC self-renewal, whereas others specify differentiated hematopoietic lineages. The bHLH factor SCL, the SET-domain containing histone methyltransferase MLL and runt-domainRunx1 factors, are essential for production, survival and self-renewal of HSCs.

MLL andRunx1 are reported to regulate HoxB4 and Pu.1, respectively, whereby the loss of either factor blocks self-renewal. Presumably, other yet unidentified downstream regulators may also be important. One lesson drawn from the identification HSC regulators is that many of these are concurrently mis-expressed in various human leukemias. For instance, SCL is associated with T-cell acute leukemia, and MLL and Runx1 is associated with myeloid and lymphoid leukemias.

The transcriptional control of genetic targets by these regulators during hematopoiesis and oncogenesis has not been forthcoming. It remains uncertain if a common factor does indeed controls similar self-renewal genes which is regulated in one cell type but deregulated in the other.

Co-factors and Post-translation Modifications Impart Transcriptional Specificity

Important principles underlying the transcriptional control of erythropoiesis have emerged from mechanistic studies on the synergism and antagonism of factors involved in lineage transition. Gata1, the founding member of the GATA TF family is a determinant for common myeloid progenitor specification. Analyses of chromatin occupancy revealed that Gata1 occupies only a small proportion of high-affinity GATA motifs in the genome. Interestingly, Gata2, which contains a nearly identical dual zinc finger DNA binding domain, function uniquely to control distinct aspects of hematopoiesis. Gata2 is expressed in HSCs to promote their proliferation, survival and function.

Intriguingly, during erythropoiesis, Gata1 and Gata2 show reciprocal expression, with the levels of Gata1 increasing and Gata2 declining. Since both Gata1 and Gata2 could occupy similar target genes containing the GATA motif, how does the subsequent expression of Gata1 determine the activation of genes necessary for erythropoiesis? This selectivity appears to be partially imparted by the cell-specific co-regulator Friend of Gata1 (FOG-1) which mediates certain functions of Gata1. Co-regulator proteins can be recruited to chromatin by sequence-specific TFs and catalyse chromatin modifications that control DNA accessibility and binding of the transcription machinery. However, the logic which dictates specific recognition of cognate Gata1 versus Gata2 targets, as well as their unique transcription output remains unclear. The question of how related factors are recruited to common versus distinct chromatin target sites needs to be resolved.

While the intrinsic differences in biochemical properties between related TFs are often under-appreciated, it could explain the phasic expression of related members which may perform unique and/or redundant functions. For example, the considerable protein stability of Gata1 compared with Gata2, conferred by the C-termini, provides an alternative explanation for selective target gene occupancy. Gata1 is subjected to phosphorylation and acetylation, but it is unclear how these modifications affect Gata1 activity or function.

The relevance of post-translation modifications to stem cell factors in defining genome-wide localization has not been examined. Although phosphorylation of the downstream effectors, Stat3 and Smad, of the Jak and BMP signaling pathways are essential for nuclear localization and DNA binding properties in ESCs, it is unclear how such modifications alter the binding dynamics of other stem cell factors on a global scale.

There is emerging evidence that ESC factors such as Oct4 is subjected to sumoylation that enhances its stability and transactivation function. The Src family of tyrosine kinases has also been reported to be important for ESC self-renewal, emphasizing the under-appreciated aspect of post-translation modifications in controlling transactivation activity and occupancy profile. Thus, differential stability of TFs can determine chromatin occupancy and help establish the appropriate genetic networks.

Surrogates to Somatic Stem Cells

Stem cells and cancer cells share several traits. Distinct from normal somatic cells which contribute to tissue function and homeostasis, both cell types are capable of sustained self-renewal in vitro and in vivo. The scarcity of cultured stem cells has, in part, led to the use of cancer cell lines such as Burkitt's lymphoma cells, colourectal cancer cells, gliomas and T-cell acute lymphoblastic leukemic cells as surrogates for genome-wide TF location analyses. For example, the immortalized human non-neuronal T-lymphocyte cell line Jurkat is used to map binding sites for REST.

Consistent with its current knowledge as a repressor of neuronal gene expression, a significant proportion of its targets is related to neurons and their development, synaptic transmission, and nervous system development. The oncogene c-Myc initially identified to be elevated in the lymph cells of patients suffering from Burkitt's lymphoma due to chromosomal translocations, has been later discovered as one of the key factors for inducing pluripotent stem cells from somatic cells. The pervasive importance of c-Myc in underlying stem cell and cancer cell functions has led to its genome-wide mapping in Burkitt's lymphoma cells and ESCs.

In both self-renewing populations, c-Myc is recruited to similar categories of genes that are involved in metabolism, cell growth and proliferation, and apoptosis. The resemblance, in which a TF may be expressed in different cell types but operate in an apparently similar context, thus provides compelling justifications for using cell lines as surrogates. However, it is imperative to note that surrogates provide at best a rough approximation of TF function. Even tissue stem cell lines arguably provide an estimation of their in vivo function owing to the effects of long-term culture adaptation and exposure to signals different from the niche context. On-going efforts in the generation of tandem affinity-tagged stem cell-specific TFs amendable to high quality immunoaffinity capture in transgenic animals constitutes the next step forward in unraveling transcriptional targets in the physiological context.

REACTIVATING PLURIPOTENCY

Somatic cells are functionally specialized cells programmed to be in a stable state. In a landmark experiment, the somatic nucleus from Xenopus can be reverted to totipotency by an enucleated oocyte, and becomes capable of recapitulating complete animal development. Subsequently, the same phenomenon was recapitulated with the mammalian sheep somatic nucleus. The reversal of a differentiated cell to a developmentally more primitive state is termed 'reprogramming'. Using an in vitro system, the generation of iPS cells from somatic cells through careful manipulation of specific molecules has been demonstrated. By over-expressing a limited but variable combination of ESC-associated molecules comprised of Oct4,Sox2, c-Myc, n-Myc, Lin28, Nanog, Klf2 and Klf4, stem cells can be induced from embryonic and adult somatic cells such as fibroblasts, keratinocytes, liver and stomach epithelial cells of mouse and human. These karyotypically normal cells exhibit morphology, growth properties and molecular signatures similar to ESCs. Notably, iPS cells can contribute to all germ layers of the chimeric mouse, and are capable of germ-line transmission. Here, an understanding of the transcription targets of these 'reprogramming factors' has provided useful insights into the mechanism for pluripotency reactivation.

In ESCs, Oct4 and Sox2 bind and regulate downstream genes, which corroboratively contribute to reverse the differentiated cell phenotype, through controlling a host of genes responsible for maintaining pluripotency or inhibiting differentiation described previously. The relevance of c-Myc and Klf4 targets appear to be more speculative. From recent studies, the role of c-Myc in sustaining self-renewal may be attributed to its many targets that enhance proliferation, negatively regulate differentiation, cellular transformation, and regulation of chromosome accessibility to other factors. The permissiveness of chromatin to TF binding and transcription activity is partly determined by histone modifications. The specific covalent modifications on the histone moieties of nucleosomes can be correlated to the transcription activity of genes in either promoting or repressing their expression. In ESCs, a large set of developmentally important genes that are silent but activated upon differentiation carries the bivalent marks, consisting of the activating histone 3 lysine 4 trimethylation (H3K4me3) and the repressive H3K27me3 modifications. This suggests lineage-specifying genes are primed for activation upon ESC differentiation.

A global assessment of the bivalent histone marks of c-Myc target promoters has revealed that the majority (95%) bear the H3K4me3 mark, suggestive of a relationship between c-Myc occupancy and histone modifications. Indeed, these target genes are more frequently expressed in ESCs, compared to other factors. The data indicates that c-Myc occupancy is associated with global epigenetic modification of the chromatin at its targets, distinct from the role of other core factors which may manifest bivalent domains comprising of both H3K4me3 and H3K27me3 marks (Kim et al., 2008). This latter data is coherent with observations in human and mouse ESCs where there is considerable overlap between H3K4me3 and H3K27me3 loci across the pluripotent genomes. In addition to histone methylation, c-Myc may also potentially modify the chromatin architecture to be more permissive for the binding of Oct4and Sox2, through its regulation of histone acetyltransferase complexes that induce global histone acetylation. The role of Klf4 in ESCs and in reprogramming is less clear, but circumstantial observations hints that it could act as a tumor-suppressor gene in iPS cells.

Klf4inhibits cell growth in mouse and can reduce the tumorigenicity of colon cancer cells in humans. This may be correlated to the observations that during the reprogramming process, bona fide iPS cells are sparsely scattered amidst large numbers of transformed colonies. Presumably, Klf4 could act as a selector for truly reprogrammed cells over transformed or incompletely reprogrammed cells which are deregulated in proliferation control. Klf4 may also act via repression of p53 which leads to the activation Nanog and other ESC-associated genes. The occupancy of Klf4 on Oct4, Sox2 and other common downstream targets that include Nanog and c-Myc has revealed that it is an upstream regulator of positive feed-forward loops that sustain high levels of the activated factors.

Interestingly, Klf2 can replace Klf4 in reprogramming and it has been shown that Klf2, Klf4 and Klf5bind to similar targets in ESCs. The amenability in which the combinatorial use of reprogramming factors results in the generation of iPS cells has raised interesting speculations pertaining to the plasticity of reprogramming. On the one hand, certain TFs appear mutually replaceable to target common genes possibly through motif recognition; on the other hand, the activation of distinct but functionally similar target genes may, in concert, results in induced pluripotency.

Further attempts to dissect the mechanism of reprogramming through examination of partially reprogrammed cells have provided strong evidence of a direct connection between transcription regulation and epigenetic modification during the transition of cellular states. Whereas fully reprogrammed cells manifest gene expression and epigenetic profiles that are similar to ESCs, partially reprogrammed cell lines do not completely repress lineage-specifying TFs and show DNA hypermethylation at the loci of pluripotency-related genes.

Strikingly, silencing of specific lineage-specifying TFs and treatment with DNA methyltransferase inhibitors help improve the efficiency of the reprogramming process, supporting genome-wide observations that methylation of CpGs are dynamic epigenetic marks that are subject to extensive changes during differentiation and need to be appropriately reversed at specific genes during reprogramming. Presumably, partially reprogrammed entities are trapped in the intermediate state as a result of either the inappropriate silencing of pluripotency factors or the mis-expression of differentiation-associated genes. A better understanding of the role of lineage-specifying TFs, and the requirements for epigenetic landscape that include DNA and histone modifications at specific gene loci, may facilitate in reprogramming differentiated cells to somatic stem cells which are clinically relevant.

Cell type-dependent Regulation of Distinct Circuitries

The molecular mechanism for the specification of embryonic germ layers and extra-embryonic lineages during pre-implantation development is an important area that is not well-understood. With genome-wide TF localization analyses, precise and detailed maps can provide a comprehensive view of the molecular circuitry governing embryonic development. Although much effort has focused on how different TFs act singly or collectively to regulate extensive networks, there is scant knowledge on how the same TF may behave in different stem or progenitor cells. This is largely due to the paucity of identified molecules that are common in more than one stem cell types, but yet retain discrete functional significance in each lineage.

In several cell types, the mapping of lineage-determining TFs has placed them at the top of transcription hierarchies. These factors are largely auto-regulatory, but also orchestrate the expression of downstream effectors. These effectors may independently or collaborate with other factors to control the cellular response. The observation that a TF controls its own expression is not surprising as auto-regulatory loop appears to be a general feature of 'master' regulators. For example, HNF TFs are required for the normal function of liver and pancreatic islets. In human hepatocytes, HNF1α binds a substantial number of genes related to hepatocyte biochemistry such as gluconeogenesis, carbohydrate synthesis, lipid metabolism and detoxification. This network also includes HNF4α and HNF6 which co-occupy the promoters of one another, thus forming a core network.

Lineage-specific Recruitment

One central theme has been the prevalent observation that a lineage-specific factor controls of a discrete transcription network. Recently, we have identified Sall4 as a key regulator common to two embryo-derived stem cell lines that are developmentally connected (accepted, Cell Stem Cell). Our deciphering of Sall4 targets is timely as integrative studies from the mapping of multiple TFs has identified Sall4 to be a major regulatory node in ESCs. Many Sall4-regulated genes are known to play important roles in maintaining ESC pluripotency. Of significance, many homeodomain protein promoters occupied by multiple factors such as Nanog, Sox2, Oct4 and Klf4, are also regulated bySall4. These findings thus integrate Sall4 as a major component of the core ESC circuitry.

In addition to defining its role in ESCs, we demonstrated the importance of Sall4 function in the primitive endoderm stem cell derivative, XEN cells. Our studies reveal Sall4 as an upstream activator of key lineage-defining genes in the primitive endoderm. Gata6 and Gata4 are two of the earliest determinants of the primitive endoderm. However, other factors appear necessary for the initial primitive endoderm specification as Gata6 and Gata4 mutants do not exhibit extra-embryonic endoderm defects until several days after blastocyst formation. The discovery that Sall4 lies upstream ofGata6 and Gata4, and drives the expression of other lineage-determining genes, provides a plausible reasoning for why the loss of either factor does not lead to a complete disruption of primitive endoderm formation, but only affects visceral endoderm differentiation at later stages. These observations thus support Sall4's role as a central player in primitive endoderm and XEN cell maintenance.

Genome-wide mapping of the same TF in two distinct stem cell lines has revealed several novel findings. Firstly, Sall4 regulates distinct genes in the different cell types. In ESCs, pluripotency-associated genes predominate; whereas in XEN cells, endoderm lineage-associated factors that includeGata4, Gata6, Sox7 and Sox17 are targets. Secondly, while the recruitment of Sall4 and its interacting partners appear necessary to coordinate the expression of a subset of genes in establishing a specific lineage programme in one cell type, it is equally important that the promoters of the same subset of genes become inaccessible to Sall4 in another stem cell type to ensure the complete absence of Sall4 regulatory input. This study illustrates and provides a basis for future investigations into how other similar factors could have unique roles in more than one progenitor cell type, as well as further our understanding on the genetic and epigenetic bases for self-renewal and lineage-potency in stem cells.

Cofactor-mediated Differential Recruitment

The interaction of a sequence-specific TF with co-factors

plays a contributory role in its differential recruitment. Sall4 physically interacts with several factors that include Nanog, Dax1, Nac1, Zfp281 andOct4 as protein complexes in ESCs. However, the identity of its partners in XEN cells is unknown. Whether the recruitment to specific promoters in different cell types is partially mediated by other sequence-specific co-factors remains to be tested. Notably, the reliance on co-factors for conferring recruitment specificity appears to be a recurrent theme that has been observed in non-stem cells. For example, FoxA1 (HNF3α), a member of the Forkhead family of winged-helix TFs, is involved in the organogenesis of liver, kidney, prostate, and mammary gland.

Interestingly, the high expression of FoxA1 in tumors arising from the prostate and breast has led to speculation on the similarities or differences of its transcription dynamics between these cell types. In breast cancer cells, FoxA1 behaves as a pioneer factor that mediates estrogen receptor α (ERα) recruitment to is-regulatory elements, whereas it interacts with the androgen receptor (AR) in prostate cancer cells.

A comparison of the FoxA1 binding loci between the breast and prostate cancer cells has shown that FoxA1 is recruited to largely different sites across the genome. Coherent with similar observations made in ESC and XEN cells, FoxA1 appears to regulate discrete transcription programmes as a result of lineage-specific recruitment. While the transcription landscape of ESC factors is primarily associated with promoter occupancy, the ERα or AR-dependent recruitment ofFoxA1 occurs at enhancers. The Forkhead motif is enriched at the binding loci both cell types, but the recognition motifs for ERα and AR are specifically enriched in FoxA1 binding loci unique to either breast or prostate cancer cells, respectively. The extensive occupancy of TFs at enhancers on a global scale supports the notion of enhanceosomes which may be prevalent features in many cell types.

The transcription switch in the regulation of differential gene sets following cellular perturbation provides an interesting framework for testing the dynamicity of factor localization to the chromatin, as well as the requirements of co-factors and/or epigenetic permissiveness. For example, $NF_{-k}B$, a key mediator of inflammation, responds to lipopolysaccharide (LPS) stimulation in monocytic cells. On a global scale, $NF_{-k}B$ recruits E2F1 to fully activate the transcription of $NF_{-k}B$ target genes upon LPS induction, whereby this effect could be abolished through the depletion of E2F1 by RNAi (Lim et al., 2007).

This fluid nature of TF recruitment raises interesting questions of whether forced expression of co-factors not naturally present in the target cell could induce the binding to, and expression of alternate genes, and elicit a change in phenotype. Through this manner, the forced ectopic expression of ESC TFs in somatic cells has been shown to dramatically alter the transcription landscape and cellular identity, leading to the induction of pluripotent cells.

THE INTEGRATION OF MICRORNAS IN TRANSCRIPTION CIRCUITS

The governance of stem cell identity is not limited to transcription networks. Post-transcriptional regulatory mechanisms modulate the expression of many genes, emphasized by the many points of control in eukaryotes. It is now apparent that a class of non-coding RNAs, microRNAs, modulates mRNA decay and translation rates. MicroRNAs are necessary for many aspects of animal development and stem cell maintenance. For instance, the miR-17?92 cluster which is highly abundant in mouse ESCs has extensive roles in the development of the heart, lungs, and immune system, as well as tumorigenesis. The skin microRNA, miR-203, represses p63 and promotes differentiation of stratified epithelial stem cells. In mouse ESCs, miR-134 and miR-1 can induce differentiation into the neuroectoderm and cardiac muscle lineage, respectively. The importance of microRNAs in stem cells is further highlighted by the observation that ESCs lacking the microRNA processing machinery comprised of either Dicer or Drosha cannot differentiate into most lineages.

Despite our increasing knowledge of microRNA functions and their regulation of target gene translation, there is a dearth of information regarding the control of their expression. Unlike most protein-coding transcription units which contain obvious transcription start sites and other initiation landmarks, microRNA promoters are not well-characterized. The occupancy of a TF on a microRNA gene is at best determined by its proximity to infer some regulatory function. In human ESCs, at least 14 microRNA loci are bound closely by Oct4, Sox2 or Nanog, where miR-137 and miR-301, are co-occupied by all three TFs. In the mouse counterpart, five microRNA loci are bound by Oct4 or Nanog, where miR-296 and miR-302 are co-occupied by both.

These two microRNAs are highly expressed in ESCs and thought to contribute towards ESC function. Interestingly, the mouse miR-302a-e cluster is the homologue of the human miR-372/3 family which is highly abundant in human ESCs, and acts as oncogenes in testicular germ cell tumors. When introduced into primary human cells, miR-372/3 can induce cellular transformation, suggesting a common role for this microRNA family in cell proliferation. Although many of these ESC-specific microRNAs appear to be under the control of key TFs, there is surprisingly little demonstration of their direct regulation.

The repressor REST is observed to bind several microRNA promoters which contain REST regulatory sequences in ESCs. One of these microRNAs, miR-21, has been previously characterized as an 'oncomir' that promotes transformation in cells by targeting tumor suppressors. Its levels also appear elevated in certain primary cancer cells. While REST regulates ESC pluripotency possibly through its silencing of developmental genes associated with the neural lineage it also blocks the expression of target microRNAs which may promote differentiation. Elevating the level of miR-21 reduced the self-renewal capacity of ESCs and it is proposed that miR-21 may potentially alter the proteins levels of its computationally predicted targets, Sox2 and Nanog, although this has yet been demonstrated. Parallel work on miR-134 has shown that it suppresses the translation of ESC factorsNanog and LRH1, thereby leading to cellular differentiation.

It would indeed be fascinating if miR-21 feeds back into the transcription circuitry by suppressing the activity of its target TFs. The transcriptional control of microRNAs and their re-integration into the circuit presents a novel facet of gene regulatory network that is currently not well-elucidated. There is emerging evidence that microRNAs are integral components of transcription circuitries, both regulating, as well as being regulated by, TFs. Computation models predict that microRNA-mediated feedback and feed-forward loops are prevalent. The TF input can either positively or negatively co-regulate a microRNA and its targets simultaneously, thereby enhancing the robustness of gene regulation and stability. This may be particularly important for reinforcing the gene expression programme of differentiated cellular states.

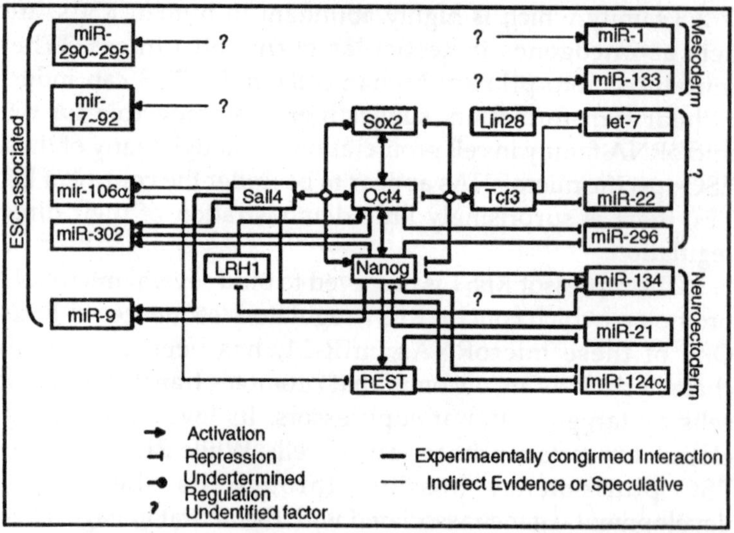

Fig. MicroRNAs are Integral to Transcription Networks in ESCs.

The circuit map that integrates transcription regulation and microRNA functions is constructed based on existing data. A core set of TFs regulates themselves, as well as the expression of microRNAs. MicroRNAs, whose expressions are positively regulated, are frequently associated with self-renewal, whereas those negatively regulated in ESCs are associated with differentiation. In addition, microRNAs feedback into the transcription circuitry through post-transcriptional modulation of their TF targets, controlling their levels and activity. Several other microRNAs appear to have roles in either self-renewal or differentiation, but their upstream regulators have yet been identified.

The recent genome-wide analyses of multiple TF occupancy could help reveal microRNAs bound by multiple factors. It is interesting to test whether the distinctive classes of factors comprising of Oct4/Sox2/Nanog and n-Myc/c-Myc/Rex1 could regulate delineating classes of microRNAs. If this is true, it would then suggest that different classes of microRNAs perform specific roles that complement their upstream regulators. From a clinical perspective, the set of microRNAs bound by reprogramming factors is potentially important to examine. It remains to be tested whether any of these microRNA targets can replace or enhance TF-mediated reprogramming of somatic cells into iPS cells. The chemical simplicity and ease of synthesis of microRNAs are key advantages in a small molecule approach to generate medically useful stem cells.

In epithelial stem cells where p63 maintains the undifferentiated state, the interaction of miR-203 and p63 in repressing 'stemness' and inducing differentiation into stratified epithelium has been shown. However, the manner in which p63 could trigger other microRNAs which help maintain undifferentiated stem cells is not known. Like transcription repressors, microRNAs are integral nodes of many gene regulatory networks. While microRNAs appear to predominantly operate through a repression mechanism, their functions in the context of networks need not be repressive. Presumably, microRNAs could suppress certain TFs, leading to the activation of a second order hierarchy that initiates the next transcription programme. The identification of microRNA signatures specific to many stem cell types such as HSCs, mammary and lung epithelial progenitor cells, and mesenchymal stem cells, has provided the opportunity to intersect the regulatory networks of transcription regulators with microRNA and their target genes.

PERSPECTIVES

Investigations into transcription regulation have progressed from the elucidation of very limited gene interaction networks to the mapping of complex extensive interactions on a genome-wide scale. This is, in part, facilitated by the employment of high-throughput technologies in detecting and analyzing TF binding sites. Knowledge of how dominant factors control gene expression in stem cells and the progression of tissue lineage commitment during development is particularly important.

As the stage-wise transition of pluripotency to multipotency and finally terminally differentiated cells requires the timely activation of a cascade of unique transcription programmes, the identification of key regulators in the hierarchy and their targets provides valuable clues into the manipulation of stem cells. Two aspects of this manipulation predominates the field of stem cell biology.

Firstly, uncovering the transcription programme of a differentiated cell type can provide information on directing the stem cell programme to proceed towards that of the desired cell type. Understanding how to mature ESCs into stable functional somatic cells is the key to generating useful cell types for transplantation.

Secondly, the derivation of patient-specific transplantable cells is a major challenge. Obtaining patient-specific ESCs has proved surmountable and has raised ethical concern. Hence, unraveling the somatic stem cell programme can aid the reprogramming of terminal differentiated cells into all the various somatic stem cell types. This can potentially bypass the need for extensive differentiation regimes that would be required for the effective use of ESCs.

The observation that certain key transcription regulators previously thought to define a particular lineage is also abundantly expressed in other cell lineages is not new. As intrinsic determinants of the cellular identity, TFs provide an entry point for uncovering how stem or progenitor cells attain their phenotype, and how lineage-specific differentiation is programmed.

However, insights into the mechanism for the differential function of these TFs are critically lacking. The simple explanation that a single factor controls a small handful of divergent lineage-restricted genes necessary for establishing distinct cellular identities does not satisfactorily resolve a complex developmental problem. The genome-wide identification of TF binding targets has brought us one step closer in our attempt to deconstruct the regulatory mechanism of stem cells.

With the elucidation of complex gene regulatory networks, the field is only starting to reveal its secrets. Much of the on-going challenges will lie in our ability to dissect the mechanistic interaction of TFs with the chromatin. In particular, how the epigenetic state determines the specificity of factor recruitment, the requirement and function of co-regulators, the dissection of a combinatorial code for transcription output, as well as the activity and biochemical properties of TFs, will be key areas to study.

Arguably, pioneering work in understanding the comprehensive stem cell regulatory network will form the basis for probing the greater depths of developmental biology, and our ontogeny.

THE PLURIPOTENT TRANSCRIPTOME

Pluripotent stem cells, such as Embryonic Stem cells, have the extraordinary ability to give rise to all cell types of the body, and hold great potential for regenerative medicine. Pluripotent stem cells are derived from in vitro culture of early embryos or by over-expression of specific transcription factors in somatic cells. In recent years great progress has been made towards understanding the molecular basis of pluripotency. The transcriptomes of various pluripotent stem cell lines have been examined in order to identify molecular mechanisms that control self-renewal and pluripotency.

These studies have begun to reveal the components and architecture of the transcriptional regulatory networks of pluripotency, and suggest that transcription factors maintain the pluripotent state in the backdrop of a loose, accessible chromatin. Such a chromatin state may allow for increased global transcription, resulting in the expression of many novel and as yet uncharacterized transcripts.

Fluctuations in the transcriptional networks may alternate pluripotent stem cells between states with distinct probabilities for self-renewal or differentiation. In this, we review studies of the transcriptome of pluripotent stem cells, highlight intriguing transcriptional similarities between pluripotent stem cells, cancer cells and the germline, and discuss future avenues of research.

INTRODUCTION

Embryonic Stem (ES) cells are the prototypical pluripotent stem cell: they are able to both self-renew and differentiate into all cell types. Because of these properties, ES cells are an excellent system to study cellular differentiation in both normal and diseased states.

In addition, cells derived from ES cells have great potential for therapeutic applications, provided they can be efficiently expanded and differentiated in vitro. An important step towards therapeutic applications of ES cells is to identify the molecules and pathways governing ES cell proliferation and differentiation.

ES cells are derived from in vitro culture of the inner cell mass (ICM) of the blastocyst, prior to implantation. Another pluripotent cell type, Embryonic Germ (EG) cells, can be derived from in vitro culture of the embryonic germline, Primordial Germ Cells (PGCs).

Recently, induced Pluripotent Stem (iPS) cells have been derived from non-embryonic somatic cells by the introduction of specific key transcription factors have uncovered gene networks and signaling pathways believed to be essential for maintenance of the pluripotent state.

Here, we review studies of the transcriptome of pluripotent stem cells, and highlight intriguing similarities between pluripotent stem cells, cancer cells and the germline.

Genome-wide expression profiling using expressed sequence tag (EST) analysis, serial analysis of gene expression

(SAGE), DNA microarrays and massively parallel signature sequencing (MPSS) have all been used to study the transcriptome of ES cells.

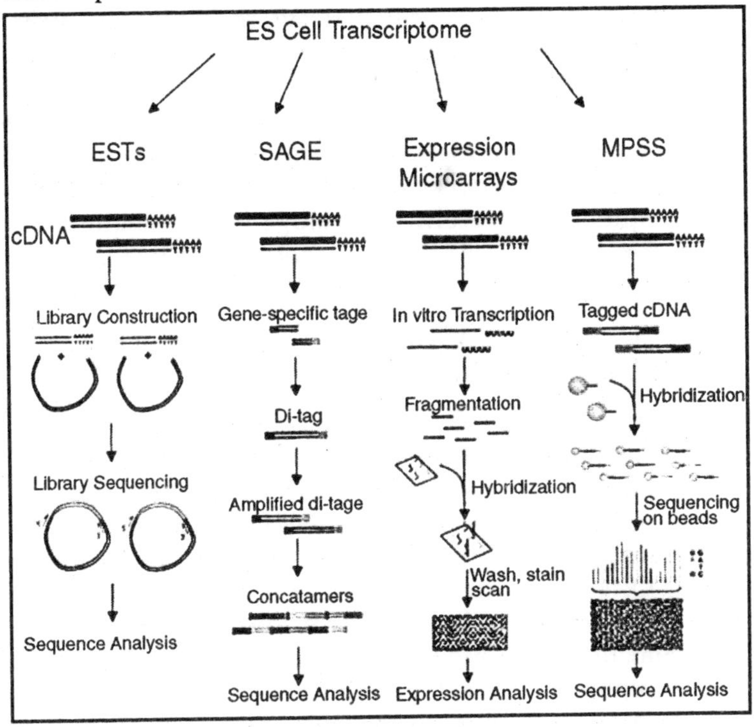

Fig. Gene Expression Profiling Techniques used in Studies of Pluripotent Cells.

ESTs are generated by inserting cDNA into cloning vectors to produce a cDNA library. The inserts are sequenced and the resulting EST sequences are analysed. SAGE involves the generation of gene-specific tags, typically 10-14 basepairs in length. These tags are ligated to form di-tags, amplified by PCR, ligated again to form concatamers and sequenced. Expression microarrays are arrays of synthesized oligonucleotides or spotted cDNAs in a predetermined spatial orientation. Total RNA is reverse transcribed, fluorescently labeled and hybridized to the microarray.

Specific hybridization signals are detected by a fluorescent scanner, and gene expression is deduced. Tiling arrays are modified microarrays where all non-repetitive genomic DNA is represented at various sequence resolutions. Unlike conventional microarrays, tiling arrays enable the discovery of novel transcribed sequences and regulatory elements through the unbiased examination of the genome. MPSS involves the cloning of a cDNA library on beads and the acquisition of 17-20-nucleotide tags from these cDNAs using an unconventional sequencing method that allows for large-scale sequencing. Recently developed deep sequencing approaches using 454 and Solexa platforms promise to provide a comprehensive description of the pluripotent transcriptome.

MOLECULAR SIGNATURE OF PLURIPOTENT STEM CELLS

To define the molecular signature of pluripotent stem cells, several groups compared the expression profiles of ES cells to those of differentiated or to ES cells of other species. An emerging picture from these studies is that some genes are predominantly or exclusively expressed in pluripotent cells, such as Oct4,Nanog, Rex1 and Utf1. These studies also point to the activity of particular signaling pathways in ES cells, such as Jak/Stat, Bmp, Wnt and FGF. Many of these studies highlight an over-representation of transcription, chromatin remodeling and DNA repair factors in the pluripotent transcriptome, suggesting 'that these play an essential role in pluripotency. Upon in vitro differentiation of ES cells, pluripotency-associated genes are consistently down-regulated regardless of the method of differentiation, while the induced genes depend on the method of differentiation.

Predominant or exclusive expression of a gene in ES cells provides insufficient evidence for a role of that gene in self-renewal or pluripotency. Some genes predominantly expressed in ES cells, such as Oct4 orSox2, had been shown to be essential regulators of pluripotency before transcriptional profiling became widely used. By combining expression profiling with RNA interference (RNAi), Ivanova and colleagues identified several novel regulators of ES cell self-renewal, such as Esrrb,

Tbx3 and Tcl1 In addition, gene expression profiling of ES cells in which expression of a critical transcription factor has been perturbed can provide insight into the transcription regulatory networks of pluripotency. For example, genes that are affected by downregulation of Oct4 are similarly affected by downregulation of Nanog or Sox2, indicating that Oct4, Sox2 and Nanog cooperate to control transcription of target genes.

Downregulation of Esrrb, Tbx3 and Tcl1 affected the expression of a different set of genes, suggesting that they regulate a distinct ES cell pathway. Thus, transcriptional profiling, in particular in the setting of gene-manipulated ES cells, will likely continue to provide important insights into the regulation of the pluripotent transcriptome.

CO-EXPRESSION TO CO-REGULATION: REGULATORY PATHWAYS

Genes co-expressed in pluripotent cells are likely to be (at least in part) co-regulated by the same transcription factors. Several key transcription factors have been identified as expressed in ES cells and involved in the maintenance of pluripotency. Among these are factors known as"core regulators of pluripotency":

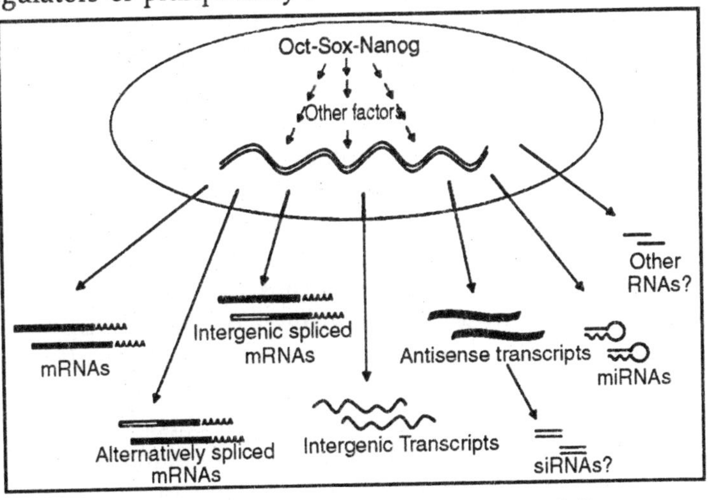

Fig. The Transcriptome of Pluripotent Cells.

The"core"regulators of pluripotency Oct4, Sox2 and Nanog, act in concert with other transcription and chromatin modifying factors to establish and maintain the transcriptional programme of pluripotency. The accessible chromatin of ES cells may allow for increased global transcription resulting in expression of many genes, their splicing isoforms, intergenic spliced mRNAs, non-coding intergenic transcripts and antisense transcripts. The function of these novel transcripts in ES cells is not yet understood. Antisense transcripts are thought to regulate the expression of their sense counterparts, possibly by sense-anti-sense pairing and production of siRNAs. Endogenous siRNAs have been recently detected in oocytes, but so far they have not been described in ES cells. miRNAs however, have been shown to be expressed in ES cells and are required for their proliferation and differentiation.

The POU-family transcription factor Oct4 is predominantly expressed in pluripotent cells of the developing mouse embryo (ICM, PGCs) and their in vitro counterparts. Embryos lacking Oct4 die at the blastocyst stage, and ES cells with perturbed levels ofOct4 expression exit self-renewal and differentiate. The Sry-related transcription factor Sox2 can act synergistically with Oct4 to activate Oct-Sox enhancers, which regulate the expression of pluripotent stem cell-specific genes, including Nanog, Oct4 and Sox2 itself. A recent report indicates that the essential function of Sox2 may be to stabilize ES cells in a pluripotent state by maintaining the appropriate level of Oct4 expression.

Similarly to Oct4, the homeobox transcription factor Nanog is predominantly expressed in ICM, PGCs, ES and EG cells. Nanog has been shown to sustain pluripotency in ICM and ES cells, and to support self-renewal of mouse ES cells in the absence of LIF, a cytokine essential for mouse ES cells. Recent genome-wide studies using chromatin immunoprecipitation combined with microarrays (ChIP-chip) or paired-end ditag sequencing (ChIP-PET) revealed Nanog as part of a transcriptional network that regulates the activity of numerous genes in collaboration with Oct4and Sox2.

Since each of these three genes contains binding sites for

each of the three transcription factors, it was suggested that they may act as a self-organizing feed-forward loop that sustains the expression levels needed to maintain pluripotency. However, Chambers et al recently showed that it is possible to derive ES cells that lack Nanog but still retain the expression of pluripotency markers, including Oct4 and Sox2. Although they are prone to differentiate, these ES cells are capable of self-renewal in the absence of Nanog, and they can contribute to all three germ layers in chimeras. Therefore, it appears that, rather than being essential for establishing pluripotency, Nanog may function to reinforce the stability of the pluripotent state.

To delineate regulatory relationships between Oct4, Sox2 and Nanog and their targets, Walker and colleagues combined available ChIP-chip data with time-course gene expression profiling of differentiating ES cells, as well as ES cells deficient for Oct4 or Sox2. They found that polycomb group genes Eed and Phc1are both bound by Oct4and Nanog, are downregulated upon downregulation of Oct4 and Nanog, and all four factors target the promoters of the same set of genes, implicating them as partners in repressing genes associated with development. While gene expression or transcription factor binding studies alone are not sufficient to identify regulatory transcription factor-target interactions, this integrative approach allowed for prediction of novel regulators and regulatory networks controlling ES cell fate, some of which were verified using RNAi combined with gene expression analysis.

To further expand the core transcriptional regulatory network of pluripotency, Kim et al. used a modified ChIP-chip to examine the function of six additional transcription factors (Klf4, c-Myc, Nac1, Zfp281, Dax1and Rex1) in mouse ES cells. They showed that regulation of target genes correlates with the extent of promoter occupancy by multiple factors: promoters bound by few factors tend to be repressed, while promoters bound by more than four factors are largely active in the pluripotent state and become repressed upon differentiation. Interestingly, while the majority of the nine factors tested share similar targets, c-Myc and Rex1 occupy promoters of a distinct subset. In addition, c-Myc target genes are associated with the active histone mark H3K4me3 and are more frequently expressed than targets of other factors in ES cells.

These results are in agreement with the proposed function of c-Myc in the widespread maintenance of active chromatin and suggest a distinct function of c-Myc in regulation of chromatin accessibility in pluripotent cells. In a similar study, Chen et al. used ChIP-seq to determine binding sites of 13 sequence-specific transcription factors (Nanog, Oct4, STAT3, Smad1, Sox2, Zfx, c-Myc, n-Myc, Klf4, Esrrb, Tcfcp2l1, E2F1 and CTCF) and 2 transcriptional regulators. They found that the majority of genes upregulated in ES cells are bound by either Nanog, Oct4, Sox2, Smad1 and STAT3, or c-Myc and n-Myc. Smad1 and STAT3 share many target sites with Nanog, Oct4 and Sox2, integrating BMP and LIF signaling pathways, respectively, with the core ES cell transcription regulatory network. These two studies demonstrate that gene clustering based on combinatorial transcription factor occupancies has the potential to predict ES-cell specific gene expression.

ChIP-chip, ChIP-PET and/or RNAi in combination with gene expression profiling have proven to be valuable methods to dissect transcriptional regulatory networks. Nevertheless, they all require prior knowledge of the relevant transcription factors. With computational approaches, however, it is possible to identify novel transcription factors using only gene expression profiling data. Genes with co-regulated expression often share cis-regulatory motifs, corresponding to transcription factor binding sites, in their promoter and enhancer regions.

These motifs include the binding sites for transcription factors Oct4, Sox2 and Myc, as well as several novel motifs. The analysis of one of the novel motifs identified transcription factor Nfy as a regulator of gene expression in ES cells that is required for their proliferation. Nfy had not previously been implicated in regulation of pluripotency. In addition, gene expression profiling failed to detect Nfy as enriched in pluripotent cells, possibly because one of the Nfy subunits, Nfya, has two splicing isoforms. During ES cell differentiation, the expression of the two isoforms changes in opposite directions, which would be overlooked if expression levels of only common exons were analysed, as is normally the case with standard microarrays. Therefore, it is important to use unbiased gene profiling approaches to obtain a more comprehensive description of the pluripotent transcriptome.

NOVEL GENES AND SPLICE VARIANTS

A limitation of gene expression studies using standard microarrays is that novel non-annotated genes may not be probed for in the array platform used. A further limitation is that standard microarrays do not generally distinguish between differentially spliced isoforms. Studies of the ES cell transcriptome that did not rely on microarrays revealed that ES cells express numerous uncharacterized or novel transcripts. A large number of ESTs and SAGE tags with no match to a known gene were found to be abundantly and specifically expressed in ES cells.

In addition, the analysis of gene trap screens revealed thousands of novel exons and genes, many of which are found in gene trapping hotspots associated with loci significantly expressed in ES cells. Further studies identified naturally occurring antisense transcripts for several pluripotency genes, including Oct4 and Nanog, suggesting that sense-antisense pairing may have a role in regulation of pluripotency. It should be noted that antisense transcription is not specific for ES cells: recent analysis of global transcription in mouse revealed that a large proportion of the genome produces transcripts from both strands.

A next step will be to characterize novel genes expressed in ES cells and evaluate their role in the regulation of pluripotent cells. To this end, Kunarso et al. have analysed several novel transcripts predominantly expressed in ES cells and found that they represent novel protein-coding genes or intergenic splicing isoforms. Using combined computational and experimental approaches, Pritsker et al. have identified alternative splicing of numerous genes in ES cells, with high levels of alternative splicing in components of signaling pathways that are functional in stem cells. While the identification of novel transcripts and alternative splicing isoforms increases the complexity of the pluripotent transcriptome, future work should reveal what role, if any, they play in maintaining the pluripotent state of ES cells.

IS EVERYTHING TRANSCRIBED? IN EVERY ES CELL

The increase in the number of expressed genes as well as a high number of novel transcripts in ES cells may be due to two possibilities: 1) the transcriptome of ES cells only appears to be more complex than that of other cell types, due to the extensive investigation of ES cells; or 2) ES cells do express a higher number of genes compared to differentiated cells. To investigate these two possibilities, Efroni et al. used tiling microarrays to compare genome-wide transcription profiles of ES cells and neuronal progenitor cells. Their analysis revealed genome-wide hypertranscription specifically in ES cells, including expression of normally silent, noncoding regions. In addition, several tissue-specific genes are expressed at low levels in ES cells, suggesting their expression might occur stochastically within the population.

Increased global transcription in ES cells is in agreement with specific features of the ES cell chromatin, such as a"loose"chromatin structure and the presence of bivalent chromatin marks. In addition, ES cells contain higher levels of acetylated H3K9 and methylated H3K4, markers of open chromatin. Therefore, it seems that the increase in global transcription in undifferentiated ES cells is an inherent characteristic of ES cells. However, it is still not clear whether this hypertranscription is simply a consequence of the chromatin properties of ES cells, or whether it has a major role in regulation of pluripotency and self-renewal of ES cells. Furthermore, it is possible that some of the hypertranscription detected in ES cells is derived from heterogeneity in the cell population.

Genome-wide transcriptional profiling typically requires

a high number of cells and has therefore been performed using populations of ES cells. However, recent reports have demonstrated cell-to-cell differences in the expression patterns of several genes among undifferentiated ES cells, suggesting that ES cells are not a homogeneous population. Indeed, Toyooka et al. have identified two populations within ES cells that differ in Rex1 expression level and havedistinct differentiation potential. Similarly, Nanog is undetectable in a subpopulation of undifferentiated ES cells. This subpopulation of cells is predisposed to differentiate but may remain undifferentiated and can subsequently re-express Nanog. These results show that ES cells are a heterogeneous cell population consisting of different types or states of cells.

It has been suggested that ES cells may exist in a metastable state which fluctuates between a Nanog-Oct4-Sox2 stabilized state, where differentiation-associated transcription is maintained below threshold levels, and a Nanog-deficient state, where an increase in specific signals, such as phosphorylated-Erk, may trigger differentiation-associated transcriptional networks. Future work should reveal if heterogeneity of undifferentiated cells is limited to the cells tested, or if it is an intrinsic property of all ES cells.

The differential expression of genes in subpopulations of ES cells may be a product of stochastic gene expression, or it may be regulated by an as yet unidentified mechanism (for example, linked to the cell cycle stage). It is also possible that only a subset of undifferentiated ES cells in culture retain full pluripotency. Future work focusing on single-cell transcriptome analysis should clarify the profile and potential of different subpopulations of ES cells.

DIFFERENCES BETWEEN MOUSE AND HUMAN ES CELLS

Pluripotent mouse and human ES cells by definition share the fundamental properties of self-renewal and the ability to give rise to cells of all three germ layers. Therefore, it is not surprising that they both express components of the core transcriptional network of pluripotency, such as Oct4, Sox2 and Nanog. Nevertheless, there are important differences in the targets of Oct4, Sox2 and Nanog between mouse and human ES cells.

These differences are reflected in the divergence between mouse and human transcriptomes, and correlate with differential cytokine and growth factor requirements, growth rates and marker expression between the ES cells of the two species.

For example, human ES cells lack transcripts of the LIF signaling pathway, while this signaling pathway is expressed, active and required for mouse ES cells. Similarly, several components of the FGF pathway are expressed in human ES cells while being absent from mouse ES cells. Cell surface markers SSEA1, SSEA3 and SSEA4 are also differently expressed in mouse and human ES cells.

The differences between mouse and human ES cells may represent species-specific transcriptional programmes. Alternatively, mouse and human ES cells may represent different stages of early development. Consistent with the latter hypothesis, pluripotent epiblast cells (EpiSCs) can be derived from post-implantation mouse embryos. Unlike mouse ES cells, but similarly to human ES cells, mouse EpiSCs require Activin/Nodal and FGF signaling for their derivation and maintenance (reviewed in and are able to differentiate into trophectoderm. In addition, there is a greater overlap between genes bound by Oct4 in human ES cells and mouse EpiSCs than between human and mouse ES cells. Therefore, it is possible that mouse and human ES cells correspond to different stages of embryonic development. Future studies of EpiSCs may reveal the extent of species- and stage-specific differences between the mouse and human pluripotent transcriptome.

In summary, there is substantial evidence that both conserved and divergent pathways act to regulate self-renewal and pluripotency in mouse and human ES cells. Even though mouse ES cells are an excellent model to study developmental processes and diseases, care must be taken when extrapolating results obtained using mouse ES cells to human ES cells.

EXPRESSION OF THE PLURIPOTENT TRANSCRIPTOME IN CANCER

Cancer cells are capable of self-renewal and possess high phenotypic plasticity, which are the major features of stem cells.

In fact, it is thought that some cancers may arise from aberrant stem or progenitor cells, or that cancer cells develop by progressive de-differentiation of their normal counterparts. Such de-differentiation might occur if a particular combination of stem cell-associated factors is expressed in cancer cells, perhaps in a manner analogous to the generation of iPS cells. Indeed, ectopic expression ofOct4 inhibits progenitor cell differentiation and causes dysplasia in epithelial tissues.

It should be noted, however, that somatic cancers have not been conclusively shown to expressOct4: the expression reported by several groups was recently attributed to the expression of an Oct4pseudogene or of Oct4B, an isoform of Oct4 that is not associated with pluripotency. Recently, two groups have analysed the expression of particular ES cell-associated gene sets in various tumors.

Interestingly, they found that the ES cell-like transcriptional programme is active in several human tumors and is correlated with metastasis and mortality. This ES cell-like transcriptional programme includes genes with roles in cell cycle, signaling, transcription, DNA repair, stress response and differentiation.

The targets of Oct4/Sox2/Nanog are found to be expressed in some cancers despite the lack of Oct4 and Nanog, indicating that the pluripotency-associated transcriptional programme may be activated by alternative mechanisms. Furthermore, Wong et al. have shown that c-Myc can activate the ES cell-like transcriptional programme in adult epithelia and in an in vivo model of human epithelial cancer, leading to an increase in the number of tumor-initiating cells. In particular, expression of c-Myc led to induction of Sox2 and repression of differentiation regulators, such as Hox genes.

These findings are in agreement with the role of c-Myc as one of the key transcription factors involved in inducing pluripotency and may relate to the function of c-Myc in the widespread maintenance of active chromatin. It should be noted, however, that pluripotent cells can be derived from somatic cells in the absence of c-Myc, albeit ith a lower efficiency.

Some germ cell tumors, which are thought to arise from transformation of PGCs, have transcriptional profiles similar to ES cells, including the expression of Oct4, Sox2 and Nanog. It is important to note that the expression of Oct4,Sox2 and Nanog is not sufficient to cause tumors: normal PGCs express all three factors at levels similar to those in ES cells.

To identify genes with possible role in germ cell tumor formation, we compared the expression profiles of PGCs to pluripotent stem cell types (ES cells, EG cells and ICM) and to publicly available expression data from germ cell tumors. We found that the genes absent from PGCs but highly expressed in pluripotent cell types are up-regulated in some germ cell tumors.

Among these is Klf4, a transcription factor that functions with c-Myc to induce pluripotency in somatic cells. It is possible that Klf4, similarly to c-Myc, activates the ES cell-like transcriptional programme and induces tumor-initiating cells. In agreement with this model, expression of Klf4 in basal keratinocytes leads to squamous epithelial dysplasia (Foster et al., 2005).

These results suggest that similar mechanisms may govern the conversion of PGCs to the pluripotent stem cell state (EGs) and germ cell tumorigenesis. Further functional studies should address the role of regulators of pluripotency in tumorigenesis. Detailed characterization of the ES cell-like transcriptional networks in cancer cells may reveal novel diagnostic markers and potential therapeutic targets.

The development of high-throughput genome-wide expression analyses has led to the identification of the ES cell transcriptome. There are still several issues that need to be addressed, such as permissive transcription, stochastic gene expression and heterogeneity in the population of undifferentiated ES cells.

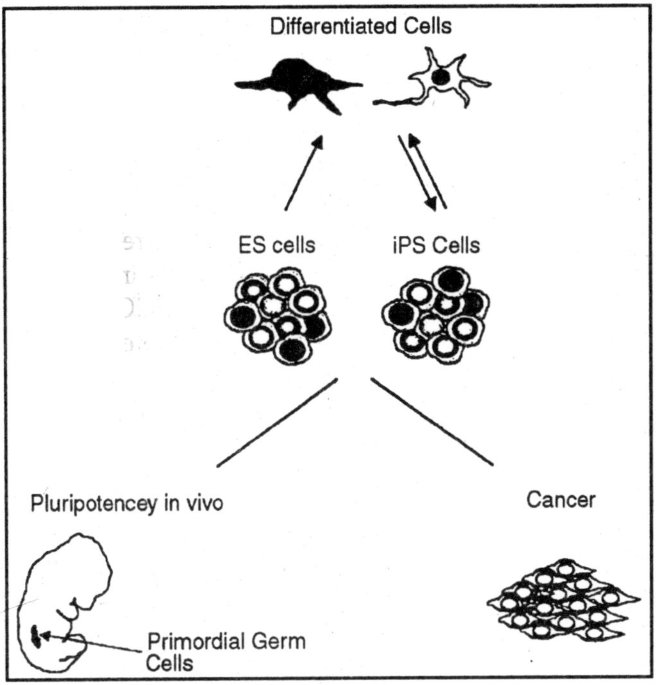

Fig. Future Avenues in Research of Pluripotent Cells.

Insights gained by analyzing the transcriptome of pluripotent cells will feed into, and feed back from, research focusing on:

- The role of pluripotency in vivo, in particular through the study of the function of regulators of pluripotency in PGCs,
- The role of ES cell-like gene expression in cancer, which may lead to development of novel diagnostic and prognostic markers as well as new therapies; and
- The regulation of ES cell self-renewal and differentiation and of somatic cell reprogramming to the pluripotent state.

IN VIVO TRANSCRIPTOMICS - A GERMLINE CONNECTION

The majority of the transcriptional profiling studies aimed

at understanding pluripotency were performed using pluripotent stem cells grown in vitro, mostly because this type of analysis requires large numbers of cells that are difficult to obtain in vivo. However, understanding the transcriptome of pluripotent cells in vivo may provide novel insights into the significance and regulation of pluripotency. Recently, we performed a comparative study of the gene expression profiles of mouse pluripotent stem cells and the cells in the embryo from which they are derived. We found that PGCs express the global transcriptional programme for pluripotency at levels similar to ES cells.

Furthermore, PGCs express high levels of pluripotency regulators such as Oct4, Sox2, Nanog, Sall4, N-Myc, Utf1 and Rex1, indicating that some of the factors and mechanisms that maintain pluripotency in early embryos and ES cells may also be operating in the embryonic germline.

The evolution of the molecular mechanisms that regulate pluripotency has not been studied in detail. The sequence and function of known regulators of pluripotency, such as Oct4 and Nanog, do not appear to be well conserved outside of mammals. Putative orthologs of Oct4 or Nanog have been identified inXenopus and chicken, but they are more broadly expressed than in the mouse embryo.

XenopusOct4 is required for maintaining the numbers of uncommitted cells during gastrulation, suggesting an ancestral role for Oct4-like proteins in preserving the undifferentiated state in early embryos. Oct4 and Nanog promote self-renewal in chicken ES cells, but their function in the chick embryo has yet to be determined. It is interesting to note that chicken Oct4 and Nanog are expressed in PGCs, and that the axolotl also contains an Oct4 gene expressed in gastrulating embryos and oocytes.

It is possible that the programme for pluripotency evolved in mammals, from molecular components already present in low vertebrates, to inhibit somatic differentiation and preserve germline competence until the germline is induced and fully developed. If that is the case, regulators of pluripotency are expected to have essential roles in PGC development.

Interestingly, Nanog-deficient PGCs fail to mature into germ cells.

As the timing of the defect corresponds to the timing when PGCs undergo extensive epigenetic changes (imprint erasure, genome-wide demethylation and X chromosome inactivation in female cells), it was suggested that Nanog functions by maintaining PGCs in a state of minimal epigenetic control, similar to the ES cell state. In addition, inactivation of Oct4 specifically in PGCs leads to a loss of germ cells. Further studies of gene expression and dissection of the transcriptional regulatory networks in PGCs should shed light on the role of pluripotency in the germline.

CONCLUDING REMARKS

It is becoming apparent that a detailed understanding of regulation of gene expression in pluripotent cells will continue to make important contributions to developmental, stem cell and cancer biology. The emerging picture is one where specific transcription factors, operating through self-enforcing mechanisms, maintain the pluripotent state in the backdrop of a loose, non-restricted chromatin. Increased global transcription in ES cells may be a by-product of loose chromatin.

Alternatively, increased transcription may be required to maintain the chromatin in accessible, pluripotent state. Fluctuations in the transcriptional regulatory network may cause ES cells to oscillate between different states with distinct propensities for self-renewal or differentiation. It will be of interest to investigate whether probabilistic behaviour is central to ES cell pluripotency, and if so how the probabilities of self-renewal or differentiation can be modulated.

Accessible chromatin and fluctuating transcriptional networks could allow for stochastic gene expression that escapes deterministic transcriptional regulation. Therefore, it will be necessary to annotate and characterize novel transcripts found in ES cells and determine whether they have a role in regulating pluripotency. The development of new, more sensitive methods, such as deep sequencing, should allow for unbiased identification of low-abundance transcripts, as well as newly emerging small RNA species. As even more sensitive large-scale analytical techniques to assess gene expression are developed, it will be important to analyse gene expression in single ES cells, and ideally to correlate individual transcriptomes to developmental potential.

The results of transcriptome analysis will have to be complemented by proteome and RNAome analysis to account for the role that post-transcriptional regulation may play in pluripotent cells. Several miRNAs were found to be predominantly expressed in ES cells, and their analysis revealed that miRNAs play crucial roles in ES cell self-renewal and differentiation. Future work will determine the role that miRNAs may have on modulating the transcriptional regulatory network of pluripotency. Finally, it will be crucial to investigate the role(s) of pluripotency in vivo in the germline and in tumor development.

Elucidation of mechanisms governing pluripotency should deepen our fundamental understanding of early embryogenesis and germline development. It will also inform our efforts to generate disease- or patient-specific pluripotent stem cells and to target differentiation towards cell types of therapeutic interest. Finally, insights gained into the regulation of pluripotency may facilitate the identification of new cancer markers and therapeutic targets. Clearly, the pluripotent transcriptome is"poised"for significant impact in a variety of fields in the years ahead.

Chapter 7

Bioethics

BIOTECHNOLOGY AND GENOME

INTRODUCTION

The genome project is the collective name for genetic mapping and sequencing efforts being pursued in various organisms by scientists around the world. These projects, primarily the Human Genome Project (HGP), arose out of past decades of genetics research, and have resulted in knowledge of the complete genetic sequence of some bacteria and yeast, and will yield that of plants and animals including human beings. The data generated is being shared by researchers around the world, but the quantity involved requires advanced genetic and computing technology.

The benefits of this grand project of molecular genetics include the rapid and economical identification of disease-related genes, and the expansion of genetic diagnosis and therapy. There will be numerous benefits for agriculture and industry. However, there are numerous ethical, legal and social impact (ELSI) issues magnified by the enormity of the project. These include privacy and genetic testing, genetic discrimination, eugenics, and genetic determinism issues. The potential economic rewards of utilisation of the gene sequences has led to questions about patenting of cDNA and other genetic data, which are still to be resolved. There has already been a shift in society during the course of this project to focus on the genetic causes of phenotype, and this trend could result in ethical abuses of the information, as well as a different view of humanity. We need to examine the role of molecular and cellular genetics in medicine and society, and educate people how to utilise genetic knowledge for the physical, mental and spiritual health of all people.

A project to study the genome of many individuals and population groups is called the Human Genetic Diversity Project (HGDP).

This raises a number of social and ethical issues including: Consent from individuals; Consent from population groups; How to obtain free and informed consent; Selection and participation; Use of the knowledge gained; Return of benefits to participants; Clash of world-views; Does the right not to know apply to communities?; Who speaks for a community?; Ownership of genes and derived knowledge; Public understanding and racism or eugenics; Stigmatization and genetic reductionism; International oversight of anthropologists and geneticists.

Population genetics is not simply a multiplication of the individual ethical and legal issues already raised by genetic research. There are different concerns and traditions in each group under study, and even among individuals within any group.

The degree of information, consultation and cooperation must reflect such differences in participants. Likewise, the role and responsibilities of the researchers and of the local and national authorities, as well as the societal implications, will differ. For the future we need consultation with populations; individual and group consent mechanisms; ongoing ethical review; inclusion of representatives of populations in decision-making; communication, education, benefits, and feedback strategies at the population level; confidential data and sample banking; continual scientific review and monitoring; and appropriate sanctions.

We also need to start gathering systematically the examples of use and abuse of DNA and cell sampling, to give us the necessary knowledge to better predict the future. Above all we need to have anthropologists and scientists who seek ethical sampling.

THE GENOME PROJECTS

The genome project is an international project aimed at obtaining a detailed map and a complete DNA sequence of the genome of a variety organisms, including mycoplasma, bacteria, e.g. Escherichia coli (completed in 1997), the yeast Saccharomyces cerevisiae (completed in 1996), the roundworm Caenorhabditis elegans, the fruitfly Drosophila melanogaster, wheat, rice, the mouse and that of Homo sapiens. It is the Human Genome Project which has captured the public attention, and has been the image responsible for increased funding of genetics research since 1988-1990.

The number of human genes that have been sequenced is exponentially growing, and it is impossible to keep a count on the progress. The expected completion date of the project is around the year 2,000 A.D. and most gene sequences are already known. The countries involved in the HGP at the time of writing include Australia, Canada, China, The European Community (especially France, Germany and the UK), Japan, Russia, South Korea and the USA. Genetic analysis of a number of other organisms is also well underway as part of the Genome Project, and the common language of genetics and similarity of many genes means that molecular and cellular genetics is being advanced by studies on all organisms.

ORIGINS OF THE GENOME PROJECT

There have been various accounts written of the origin of the Human Genome Project. The other Genome Projects arose in terminology at the same time or after the phrase"Human Genome Project"had been coined. However, all of these projects involve genetic mapping and sequencing which was underway long before the trendy phrase of"Genome Project"was used, and specific long term projects to entirely sequence some viruses, eucaryote organelle and bacterial genomes have been underway since the late 1970s.

Mapping of the human genome has been progressing for decades. The beginnings of the genome project can be at least traced back to Mendel's genetic studies on peas, the mapping of the trait for colour blindness to the X-chromosome of

Drosophila by T.H. Morgan and workers, to Avery and colleagues that found DNA was the physical substance of genes, to Crick, Franklin, Watson and Wilkins who determined the structure of DNA, to those who discovered the genetic code, to Sanger and others who developed DNA sequencing, and to many others who contributed to our knowledge of genetics and molecular biology. Therefore, no single person can claim to have initiated the goals of the genome project; and the date of the origin is obscure.

The progress in genetics since the 1960s has been rapid, and by the 1980s the goal of mapping and sequencing the complete genome of particular organisms became a distant goal. The first organisms to have their DNA sequenced were viruses, then the complete mitochondrial DNA of some organisms was sequenced.

Projects in Europe, Japan and North America began which were aimed at automating DNA sequencers in preparation for the project. In scientific conferences in the 1980s people discused the idea of a human genome project. In 1988 the Polymerase Chain Reaction (PCR) was devised, which increased the rapidity of DNA manipulation enormously. By 1992 the project had resulted in physical genetic maps of human chromosomes 21 and Y, a refined complete genetic map of human DNA; and the complete sequence of a yeast chromosome.

The Genome Project has political dimensions as could be expected for any project requiring much funding, with the huge potential rewards for biotechnology industry from the sequencing and future application of genes to medicine and agriculture. The country putting forward the most money for the project is the USA, and some would say that the genome project as a specific image arose in the USA. There were several people in the USA who saw the goals of the genome project as ideal for initiating the first large scale biological research project with a definite endpoint. The Human Genome Project or Human Genome Initiative is the collective name for several projects begun in the late 1980s in several countries, following the USA Department of Energy (DOE) decision to create an ordered set of DNA segments from known chromosomal locations, to develop new computational methods for analyzing genetic map and DNA sequence data, and to develop new techniques and instruments for detecting and analyzing DNA. The DOE has had a long term interest in assessing the effects of radiation on human health, and sequencing the genome may allow detection of mutations in the DNA. Whether the motive was to fill vacant DOE Laboratories; to provide renewed emphasis for science; or to put US biotechnology companies in a better international position, the idea itself was sure to catch the imagination of politicians.

Some biologists commented that they do not think physicists can do good biology, so the project should not be left to the DOE, and because the National Institutes of Health (NIH) is the major funder of U.S. biomedical research, the NIH joined the project. The activities of the NIH and DOE have been coordinated for genome research since they signed a Memorandum of Understanding.

The genome project is often compared to the Apollo space project. The analogy shows the glamour of the project. Not only will the genome project lead to the development of useful new technology, but unlike the Apollo project, the goal itself is also of immense direct practical use. The importance of the project initiation to reaching the goal is also different, people would not have gone to the moon if a positive decision had not been made, but the human genome map will be obtained, with or without a positive effort, though over a longer time scale if undirected. The original time frame was fifteen years, which has since been considerably shortened.

There have been numerous scientists who have contributed to our knowledge, and it will also be fitting that to complete this project will require the joint collaborate work of innumerous scientists, internationally. In the perspective of who initiated the project, who does the work, and whose knowledge is needed, the answer is clearly that many people are, and will be, directly responsible for the mapping and sequencing, and the later interpretation of the data.

GENETIC MAPPING

The DNA that is being sequenced is a composite of different human tissue cell lines. About 0.3-0.5% of the nucleotides in our DNA vary between different people. These differences vary from person to person, therefore it does not matter whose genome is actually sequenced. Different laboratories often use different human tissue culture cell lines, however, by the characterisation of standardised marker regions the DNA between different individuals will be able to be compared, and a single general map and sequence produced.

A detailed genetic map of the human genome was refined prior to full scale sequencing. A map is essential to efficient sequencing so that a physical library of DNA fragments can be systematically sequenced. The distance unit used in gene mapping is called a centi-Morgan (cM), and one cM is equivalent to two markers being separated from each other in chromosome crossing over in normal reproduction 1% of the time. The actual physical length of 1cM varies, being approximately 1 million base pairs (Mb) in humans.

The construction of any map requires markers. The markers of sequence are specific DNA sequences, and the longer the sequence the less copies there will be in any genome. A sequence of 20 nucleotides, for example, is generally unique in a genome. There are two types of maps, maps that are drawn in picture, and physical maps which include a physical collection of sequences of known order and location. Genome analysis involves both types. For short sequences, such as a few genes, convenient markers are the specific nucleotide sequences that are cut by restriction endonuclease enzymes.

A sequence of DNA can be cut into smaller specific pieces by digestion with restriction enzymes, and the pattern from digestion with several different enzymes can be used to make a map of that sequence. The use of restriction fragment linkage patterns (RFLPs) in combination with genetic linkage analysis allowed the construction of linkage maps for each human chromosome with an average spacing of 10-15cM in 1987.

The current mapping paradigm is based on a proposal in

1989 to use physical sequence-tagged sites (STS) as the map labels. These sequences are longer than those utilised by restriction enzymes, and may be 100-1000 nucleotides long, so that a single tagged site in a genome can be defined by this short nucleotide sequence. Different researchers use different cloning vectors for gene analysis, so the exchange of DNA pieces in these different vectors, or DNA clones, is not possible. What is possible using the DNA polymerase chain reaction (PCR) is to generate DNA sequences for any DNA if short sequences are known for each STS, from which primers for the PCR can be made. Therefore, the ends of large DNA fragments should be sequenced, and the data combined to make a STS map of each chromosome.

This approach means that researchers can continue to use different methods, and develop better procedures, while the information obtained can be integrated to progress the actual physical map. This will avoid the need to exchange different clones of DNA between laboratories, because each laboratory can use the marker sequence as a starting point.

The initial approach was for different research groups to concentrate on different human chromosomes in order that they could all have the complete map in a shorter time. There are actually 24 chromosomes to be sequenced, 22 autosomes and the X and Y chromosomes. An initial goal is a map with STS markers spaced at about 100,000 base pairs, and the assembly of overlapping contiguous cloned sequences (called contigs) of about 2 million base pairs length. From this physical and informational library system, the sequencing was started. Data management technology must also improve, such as programmes to search the DNA sequence libraries, using advanced computing technology.

Physical chromosome maps (overlapping cloned pieces of the complete chromosome) are made by cloning overlapping pieces of DNA, and tags on each piece allow the pieces to be put into order. This means that once a disease-causing gene is localised to a particular region of the DNA, the appropriate clone can be taken from a freezer and sequenced. In 1992, researchers in France describe improved techniques which meant that a whole human genome approach, not individual chromosome approach, could be used to construct a physical library of contigs. A French project, Généthon, took the lead in genome mapping in 1992-3 because it was the world's largest gene mapping lab. They completed the first physical human genetic map in 1993, using automatic robotic analysis systems. The library of overlapping clones is basically completed using yeast artificial chromosomes (YACs). There are minor gaps where the DNA sequences are difficult to clone. Following its rapid progress the USA began setting up two large genome mapping centres, as the approach begun to be directed at the whole genome rather than just individual chromosomes.

Using the STS approach allows small teams of researchers to contribute results. There are worldwide efforts, although the major international effort is centred around the USA/ Canada, Europe and Japan. There are many people from Australia, and Latin America who are also involved in work, via access to international DNA databanks. There will be contributions from many other countries for particular genome projects that have regional economic importance also, for example, China has a rice genome project.

Smaller genomes are useful models for the human genome, and several organisms are models for the sequencing project. The 5 Mb Escherichia coli and 15 Mb S. cerevisiae (yeast) sequences were major successes.XXX Yeast was the first eucaryote organism to have a fully sequenced genome. The first multicellular organism that is expected to be fully sequenced is the roundworm Caenorhabditis elegans. This genome project is being undertaken at the MRC Laboratory of Molecular Biology in Cambridge, U.K., and at Washington State University, USA, with other international collaborators. It is being used as a model project as a prelude to the sequencing of the human genome. The total sequence is about 100 Mb. This worm is one of the best characterised animals in the world, with the complete pattern of cell divisions for the whole organism (959 nuclei) and the full connections of the 302 nerve cells known. The life cycle is 3.5 days, which makes it easy to study mutations. The work on the physical map began in the early 1980s, and the genome is arranged in less than 50 contigs (pieces of contiguous cloned DNA sequence). In 1992 about 1 Mb of sequence was produced, and this figure is likely to increase to about 5 Mb of sequence in 1993. The cost of sequencing per nucleotide has fallen to about US 50 cents, a tenth of the cost in 1988 when the genome project began.

The model plant organism is Arabidopsis thaliana. Another model organism that has much historical significance to genetics is the fruitfly, Drosophila. The size of the D. melanogaster genome is estimated to be 165 Mb, with about 5-15,000 genes. A physical cloned map has been completed, and the sequence may also be completed before the human sequence. Drosophila has been used in mutation studies of gene function for many decades, and because of a short generation time will continue to be a major research organism for genetics. The genetic sequencing will allow detailed genetic mutation studies at a nucleotide level, which allows studies of the affects of chemical mutagens and ionising radiation.

The mouse genome is the same size as that of humans, yet there is also a mouse genome project because of the use of mice in genetics research. Most of the mouse project is underway in the USA, and a 1 cM resolution genetic map is expected by 1995. There is much similarity between the mouse and human genomes, so the maps and sequence markers share many similarities. Cattle breeders have also begun a cow genome project, pig breeders a pig genome project, and behavioural geneticists have begun a dog genome mapping project to examine genes that are linked to the selected behavioural traits of inbred dog strains.

GENETIC SEQUENCING

For many purposes there does not need to be any repository of DNA pieces, what is required is a computer data bank of the sequence. The project requires the establishment and constant improvement of databases containing the sequences of genes, and their location. There are several international databases, and the information should be openly shared among them to make the best and most up-to-date database possible. However, the physical cloned genes may be more efficiently obtained from cloned libraries, and such cloned libraries are becoming available from both the public and private sector. For example, the YAC contig library made at Généthon has been copied for several laboratories and is intended to be openly available to all peoples.

The total sequencing project for a genome of 3000 Mb will take a few years work by several large laboratories, however, by sequencing cDNA, which is complementary DNA to mRNA, was possible more quickly and cheaply. cDNA represents the expressed genes of a tissue or organism that the mRNA was obtained from. Short sequences of a few hundred nucleotides from a cDNA library were automatically sequenced by researchers in The Institute of Genomic Research (and others),, and these are called expressed sequence tags (ESTs).

The sequences are tags because they can be used to identify a gene sequence, and as a probe from which the full gene can be cloned. In 1992 the US$70 million The Institute of Genomic Research (TIGR) was founded in the USA, which is directed by Dr C. Venter. The human cDNA sequencing was largely completed by 1994. The database was sold to Smith-Kline-Beecham, and had restricted access until April 1997 when it was made open to all. In the meantime others, especially Merck Inc. had obtained separate EST databases.

FUNDING

Biomedical research is performed and funded in many countries. For many years science has not only been the individual pursuit of people with unusual ideas, but rather it involves the funding of research by public or private money, consisting of taxes, charities and business investments. We should not be surprised therefore to hear the justifications for the funding of the project in terms of the business opportunities. The US Congress was partly convinced of the usefulness of funding the project by the opportunity to boost US biotechnology. However, in France, much of the money for the genome project comes from charity, which has different implications for the way that the data will be shared, as will be discussed later.

The U.S. portion of the project (possibly 50% of the total) was initially estimated to cost US$3 billion over the next 15 years, to the intended completion date in 2005 A.D. However, new technology suggests the date will be the year 2,000, and the cost may be cheaper. The total cost is unknown because the project is being broadened to include other organisms as models, private companies are also spending money, and many countries are contributing additional money to it. Data handling will be another important portion of the total.

The 1992 government funding in the USA specifically for the human genome project (not including other indirect research on genetic mapping) was US$164 million (NIH US$105 million/DOE US$59 million). When one compares this with the cost of the development of a single drug, at US$100 million at that time (now US$300 million), or the annual U.S. health care expenditure of about trillion dollars, it is a small price to pay for such a large amount of information. Biotechnology is very big business, and the projected average US$200 million annually for the human genome project in the USA (National Research Council, 1988) is minor. The scientific methodology for sequencing DNA is routine, and the initial costs of US$1-5 for each nucleotide were reduced to 5 cents per base before the major sequencing effort began in 1996.

The initial fear of scientists was that the money would come from other biological research, which traditionally involves many small projects encouraging many individual scientists. Critics would prefer the project to stop after generating a general genetic map of the chromosomes, and after the cDNA had been sequenced. Identifying a particular disease-causing gene can take several years of intense investigation, as seen in the tracing of the cystic fibrosis gene. It has been estimated that it cost about US$30 million to trace this gene, yet if a map had existed the cost may have been only US$200,000. A more detailed map decreases the amount of DNA that must be searched through for each gene. Various leading scientists have called for an evaluation of the project priorities after the map is completed, to use the money in the best way to encourage research. This also means that the first regions of the human genome to be sequenced will be ones that are known to have disease-associated genes present. In the initial phase some of the funds for the DOE and NIH projects came from existing research funding. The NIH is spending equivalent to 2-3% of its total budget on the genome project, which in view of the importance of a coordinated mapping project, is worth the cost.

Given the potential direct medical benefits of the project, the research money that is being spent on this project, and possibly more, can be ethically justified from the principle of beneficence. From the ethical principle of justice, other countries who can afford to pay and will benefit should share the cost of the project. The reasons for more countries to join in the project is not entirely because of their recognition of the principle of justice, but includes more the elements of potential economic benefit, and fear of being denied access to the databases.

In 1993 the US funding for the human genome project from government agencies was about US$170 million, with about US$80 million from industry. In 1992 eight genome-related companies were established, as the project became more commercial. There are private companies embarking on the project. Private companies should perform such research and be able to recover costs if they are still more economical than government research, providing the results of mapping and sequencing are openly accessible. In practise much of the project will be publicly funded, but contracts to do the work may be awarded to whoever is the most competitive.

In France two private nonprofit groups (Centre d'Etude des Polymorphismes Humain and Généthon) are spending more money than the French Government on genome mapping, in pursuit of genes for humanitarian reasons, rather than economic ones. The role of nonprofit private organisations is also very important in biomedical research in other countries, leaving less room for economically motivated private companies. Time magazine has said that the French genome project is on higher moral ground than the US project, because it is donating the results to the United Nations and is not seeking gene patents.

DATA-SHARING AND PATENTING

The international mapping and sequencing is being coordinated by the Human Genome Organisation (HUGO). Coordination of the international effort is needed to avoid duplicity of effort. The European Commission is also coordinating research in Western Europe, and UNESCO has interests in coordination of worldwide research. Although people recognise the importance of mapping, many researchers remain more interested in pursuing specific disease-causing genes. Therefore more attention has been paid to chromosomes 21, 7, and X, which have unknown disease-causing genes.

A threat to data-sharing came following applications by the NIH for patents on cDNA sequences of several thousand genes, before their function was known. The US Patent Office rejected the first cDNA patent applications on all three grounds needed for patents: novelty, non-obviousness, and utility. They said that they could find some partial sequences in existing databases so the application lacked non-obviousness; and they also said it lacked novelty since they used a publicly available cDNA library. Several U.S. companies have also filed patent applications on cDNA fragments and one Japanese company has applied for patents on 60 partial sequences of full-length cDNA clones which have had their gene function inferred. Varying points of view have been expressed by patent lawyers.

There is the already existing problem of data-sharing from the viewpoint of individual competing scientists. The new technology, for example automated DNA synthesisers and the PCR, makes replication of results very rapid, which could encourage researchers to delay publication while they get more of a head start in the next stage of the research. In a system where academic jobs depend on the number of publications everyone wants to get papers published. The self-interests must be considered for the sake of the researchers' autonomy and their future work.

Researchers may not reply to letters requesting data, or just reply within a selected peer group. The U.S. NIH and DOE guidelines stipulate that data and materials must be made publicly available within 6 months of generation, and earlier if possible. The guidelines are rules for those receiving grants from the NIH or DOE genome project, but encourage researchers to share information, to avoid bureaucracy. The results of discovering a disease-causing gene, or a detailed map which advances the discovery of such genes by many years must outweigh the short-term interests of individual scientists.

An example of the collaboration is between the 30 research teams working on chromosome 21 (it contains the Down's syndrome and Alzheimer genes), who produced a physical contig map. In this respect the genome map may be the ultimate collaborative research project. The most rapid progress will come from immediate data-sharing, and it is a chance for"scientific altruism"on a global scale.

There is an ethical obligation on researchers, especially those using public money, to share data as soon as it is available. However, the financial size of the biotechnology industry that stems from molecular genetics is large enough to have already affected the way that data is shared. The net effect has probably been positive to this time, because of the large share of genetics research that is being performed in biotechnology and pharmaceutical companies.

ETHICS AND LAW OF PATENTING DNA

There are two basic approaches to applying patent law to biotechnology inventions. In the USA, Australia, and many other countries, the normal patentibility criteria shall apply, that is, the invention has the attributes of novelty, non-obviousness, and utility, and the invention should be deposited in a recognised depository. While a country may accept the first type of criteria, some countries have specifically excluded certain types of invention.

What is ethical is not the same as what is legal, though

we can attempt to reduce the difference. Concerns over ethics have affected patent laws, for example European countries who joined the European Patent Convention have barred the patenting of plants or animals. Denmark has an even stronger worded exclusion in its national law. There is public rejection of the idea of patenting animals in some countries, and the patenting of human genetic material is potentially more contentious.

To qualify for a patent an invention must be novel, non-obvious and useful. If the claimed invention is the next, most logical step which is clear to workers in that field, than it cannot be inventive in the patent sense. If a protein sequence is known, than the DNA sequences that code for it will not in general be patentable, unless there is a sequence which is particularly advantageous, and there is no obvious reason to have selected this sequence from the other sequences that code for the protein. In the case of natural products there are often difficulties because many groups may have published progressive details of a molecule or sequence, so it may have lost its novelty and nonobviousness. These are essentially short pieces of the human genome.

There are also patents on protein molecules which have medical uses. In this case the protein structure is patentable if it, or the useful activity, was novel when the patent was applied for. The invention must also be commercially useful. There are patents on short oligonucleotide probes used in genetic screening. If someone can demonstrate a use for a larger piece of DNA than they can theoretically obtain a patent on it. An example of a larger patentable section of genetic material would be a series of genetic markers spread at convenient locations along a chromosome. Another set of genetic markers on the same chromosome can be separately patented if they also meet those criteria. The direct use of products, such as therapeutic proteins, is well established. The genetic information can also be used to cure a disease, for example using the technique of gene therapy with a specific gene vector.

With the completion of the genome sequence of many

organisms, including humans, most genetic material will no longer be novel as it will be available in a database. The completion of the genome maps and sequences of many organisms will have many implications for the future of biotechnology patents.

Patenting has become an issue in population genetics primarily in relation to the patenting of products derived from the genetic material of indigenous peoples. In 1993, a patent filed by the United States government on the cell line of a 26 year old Guaymi Indian woman from Panama was opposed by the Guaymi General Congress, the World Council of Indigenous Peoples, the Rural Advancement Foundation International, and the World Council of Churches. The patent claim was subsequently withdrawn, but on March, genetic material isolated from a man of the Hagahai people from Papua New Guinea's remote highlands was patented in the USA. The NIH withdraw its claim at the end of 1996, but after provoking much controversy and distrust of scientists.

These patent applications have served to cast deep suspicion on the motivation behind human population genetics research in general. Although the primary aim of most researchers is the pursuit of knowledge, and not commercial gain, and scientists with other motives may be excluded from particular projects as the HGDP maintains, nevertheless, the possibility is that products derived from genetic material collected in population genetics research could be patented for commercial purposes. Moreover, as in the case of Moore vs Regents of the University of California (1990) where the Supreme Court of California ruled that John Moore does not have property rights in the cells taken from his body, the people who take part in population genetics research may stand to gain nothing from whatever patents that are granted on products derived from their genetic material.

The public attitudes to the patenting of human genetic material is rather negative. The negative reaction reflects the general feeling that genetic material is special, and should be different to other types of information. Patenting is said to reward innovation, which is a basis of the successful modern democratic and Asian economic systems. They do recognise property rights in inventions. However, there is an existing difference in the protection of property rights compared with other rights in international law and declarations of human rights.

Property rights are not absolutely protected in any society because of the principle of justice, for the sake of"public interest","social need", and"public utility", societies can confiscate property. Therefore there is an existing precedent for exemption from property ownership, which is the point of the exclusiveness of patents, when some property is of great benefit to the public.

Using a more positive argument, the knowledge gained should be considered as the common property of humanity. There is an existing legal concept that things which are of international interest of such a scale should become the cultural property of all humanity. It can be argued that the genome, being common to all people, has shared ownership, is a shared asset, and therefore the maps and sequence should be open to all. All people can say that the sequence is 99% similar to their own.

In the United Nations Declaration of Human Rights, Article there are two basic commitments that many countries in the world have agreed to observe (in their regional versions of this declaration). These are (italics added for emphasis) (1) Everyone has the right freely to participate in the cultural life of the community, to enjoy the arts and to share in scientific advancement and its benefits. (2) Everyone has the right to the protection of the moral and material interests resulting from any scientific, literary or artistic production of which he is the author. An important question arising from section (2) is whether all people are the author of information that is shared by all people? The writers of this declaration may not have considered DNA, but it would certainly be in the spirit of the Declaration to interpret the DNA sequence as shared ownership. In section (1), everyone has the right to freely share in scientific advancement. This article expresses two important and relevant guiding statements of law that reflect the ethical principles of justice and beneficence, and the idea of legal property rights.

The common claims for authorship of the genome should be considered in all aspects of the genome project, especially in the questions of who should make the decisions in the project and the use of data. Some of the common factors that derive from the shared ownership are that the utilisation must be peaceful, access should be equally open to all while respecting the rights of others, and the common welfare should be promoted.

The authorship of the genome can be answered in two ways. The DNA could be viewed as a random sequence of bases, and the author is the sequencer, but this is not what we would normally talk of as an author or inventor, rather the sequencers are discoverers. The sequencers of DNA are not sequencing un-owned land but rather they are sequencing un-characterised land, the name of mappers is rather suitable for this analogy. The DNA is not random, it is merely unknown. Only the method for sequencing, or mapping, can be invented and patented, and whether that side of the project can be ethically patented lies more with the question of benefit and utility as discussed above.

If these arguments are insufficient to dissuade the private ownership of genome data, in addition to the precedents for exclusion of patents, public opinion could force a policy change over the patenting of such genetic material. In the USA the commercialisation of human cells and tissues is generally permissible unless it represents a strong offence to public sensitivity. The sale of the human genome map and sequence data may be a strong offence to many and incite adverse public reaction forcing legislators to exclude it from patenting. The debate will continue, as companies will naturally desire to obtain some information protection for their investment, but they will have to be sensitive to strong public feelings that could easily be aroused, which as argued, has an ethical backdrop.

The idea that the human genome sequence should be public trust and therefore not subjected to copyright is the position of the French government, and was also the conclusion of the U.S. National Research Council (1989), and the American Society of Human Genetics. The European Parliament"Human Genome Analysis"programme limits contracting parties' commercial gains with the phrase"there shall be no right to exploit on an exclusive basis any property rights in respect of human DNA".

This idea would also include the option that the donor of genetic information, in terms of a cell line, should be able to make that information publicly available, which is usually a reasonable interpretation of the motives for patients to provide material for medical research, the motive to aid humanity in general rather than a commercial interest. The European Parliament 1993 patent rule may allow"farmer's privilege", so that farmers pay royalties on a patented gene in a seed or livestock only once, not every time the gene is passed to a descendant. It will also forbid the patenting of cloned"human parts".

Some human genes and proteins that have medical use are patented, and this type of patent has brought about huge profits for some proteins. The annual market for the protein erythropoietin in the world is over US$500 million, and some similarly profitable proteins will be found in the future. Some awards are required for promotion of the biotechnology industry and medicine. However, many people object when a routine medical technique is withheld from people who cannot afford to pay for it. Disease-related genes will be sequenced by the project, and some patents may be issued on screening methods and genes. Patent fees are being requested on the cystic fibrosis gene, even though the patent is not yet awarded. For some researchers, such patent fees may discourage studies. European researchers are challenging these royalties, and we can expect this issue to be as controversial as the cDNA patents discussed above.

BENEFITS

There are major applications and implications of the genome project for much more than molecular and cellular genetics. It will be a huge resource of information for medicine in the next century. There will be much useful information arising prior to the completion of the project, as growing numbers of disease causing and susceptibility genes are sequenced and the mutations characterised. Most of the major single gene disorders and some of the genes involved in complex diseases should be known within the decade. It will be possible to develop DNA probes to diagnose any known genetic disorder and also will be easier to characterise new disorders.

It will expand the number of human proteins that can be made by genetically-modified organisms, which would allow conventional symptomatic therapy for many more diseases, which could be supplemented by somatic cell gene therapy when appropriate. It would also expand our basic knowledge of human biology, which allows medical treatments to be developed. We may not be able to predict when therapies will emerge after the genes are discovered, because there can still be a long delay in clinical applications following biochemical understanding. The amount of new knowledge is hard for us to comprehend, it will take decades to process it all, but it offers the potential understanding of all genetic diseases sometime during the next century.

GENETIC SCREENING

Every person has a different genetic sequence, except for identical twins. Genetic screening involves the use of the binding between the nucleotides A-T and G-C. A sample of DNA is taken from a cell, and then the DNA is split into single chains. The bases in this single-stranded DNA will bind to the complementary bases of a nucleotide probe. The presence or absence of particular DNA sequences that represent different genes or mutations can be screened for. There are many types of genetic disease, some minor and others untreatable and fatal. One of the most common genetic diseases is inherited forms of cancer.

Prenatal genetic screening involves screening of the fetus, inside the mother's womb. There are several safe and commonly used methods to allow samples to be taken from the fetus. In addition to noninvasive ultrasound imaging, and the invasive sampling methods of amniocentesis and chorion villi sampling, new methods are being developed.

Two other techniques that are expected to have greater impact in the future, are isolation of fetal cells from maternal blood, and preimplantation diagnosis. Isolating fetal cells from maternal blood is advancing rapidly, and has been used for diseases such as sickle cell anemia and thalassemia. If the number of false positives found in these tests, which is non-checked could mean that an unaffected fetus is aborted, is reduced and the rate of false negatives made very low, we can expect this to become mainstream and offered to all pregnant women.

Preimplantation screening on the other hand has the advantage of selection before implantation but must be associated with embryo transfer afterwards, which requires a higher level of technical commitment and cost. Enough babies have been born to confirm its initial safety, and it is certainly less traumatic than abortion after prenatal diagnosis, at the earliest at 6-8 weeks after conception. However, it might never be widely provided, as it is limited to infertility clinics, and to parents that know that they carry a disease. It might be more ethically acceptable to using abortion, so the spirit of parental autonomy would support the provision of such services even when prenatal diagnosis is available, but it is unlikely that we could ethically justify the government funding such tests unless they provide services for general in vitro fertilization and embryo transfer for infertile couples.

Prenatal genetic screening of the fetus can be used to detect characteristics of the fetus, and should only be performed for serious diseases. These include diseases that would result in serious mental or physical health damage to the fetus. In most cases the fetus is found not to be afflicted, thus removing much anxiety from the parent's mind. Without the use of such tests to confirm the absence of disease, some mothers would have an abortion. In the case that the fetus is found to be afflicted from a disease there are two different possible courses of action. If there is a therapy available and treatment before birth is necessary to avoid health damage, the fetus can be treated for the detected disease.

There have been increasing numbers of operations performed on fetuses in the womb (in utero), which have avoided permanent health damage to the fetus. Sometimes the fetus is removed from the womb, operated on, and then replaced to complete normal gestation. If the detected disease is untreatable, and is serious, then the option of selective abortion is available.

There is considerable variability in the abortion laws of different countries around the world, but in general selective abortion of fetuses suffering from a genetic disease is a respected moral choice of the mother. There is clear public support for prenatal genetic screening in most countries that have been surveyed, and that it also should be available under government funding. There does need to be control over the use of cosmetic screening and therapy (that has no compelling medical reason), especially when it affects children. The human genome project raises similar ethical and legal issues to those in current genetic screening, such as confidentiality of the results. However, it will lead to screening on a huge scale, for many disease traits and susceptibility to disease. It is important that we deal satisfactorily with the test cases, before we are faced with all these new information. The technology may change the way we think. The amount of information obtained will overwhelm existing genetics services, and geneticists. More training of genetics (as well as ethics) will be required for health care workers, scientists, and the general public.

Will society allow individuals to have free choice over the use of genetic manipulation and screening when there is no medical reason for it? Screening can be used both to reduce individual differences to others, and to highlight differences. The arguments used against genetic intervention which has no therapeutic value include: it would be a waste of resources, may present risks to offspring, it will promote a bad family attitude, will be harmful support of society's prejudices and may reduce social variability. It will probably not have any significant affect on genetic variability as there will be plenty of alternative healthy alleles.

There could also be the idea of a natural genetic autonomy, that we should let the genes come together naturally, and let the individuals develop their genetic potential without unnecessary interference by parents or society. A criteria for transgenerational ethics is that not only must a gene alteration be safe, but it must be good therapeutic sense over many generations. There must be unquestionable objectives and benefits, for many generations.

A common feature of many issues raised by the human genome project data is that we need to consider the effects of knowledge and technology on future generations. The beneficiaries and those at risk may not yet be existing. In the sense of benefits and risks, it is their genome project more than ours. We have an obligation to the future from the principle of justice. We need to ensure future generations retain the same power over their destiny as we do, while benefiting from the culture and technology we have developed. Society's interests not only should transcend proprietary rights, but the special nature of the genome project and the claims that we can all make upon the genome should make the shared authorship and ownership legally compelling. That is why UNESCO has drafted a Declaration on Protection of the Human genome.

PRIVACY AND DISCRIMINATION

The information about whether an individual has a particular DNA sequence and gene, can be very powerful, especially in the diagnosis of genetic disease. It is very important that privacy is respected, because the information in a person's genes identifies some of our risk to disease that medical insurance companies and employers could use to discriminate people. There are already cases of discrimination of individuals after genetic testing in North America.

There is less experience with presymptomatic genetic testing of disease risk, such as for Huntington's disease. This is another area where the data needs to accumulate before we will be able to make reasonable predictions about the more widespread use of such testing. DNA sequences can be used to identify individuals, and have been used for forensic cases. When the DNA is specifically cut up by restriction enzymes and the pieces are separated unique patterns are made, called DNA fingerprints. They have been used for hundreds of court cases around the world. In several countries a genetic register of criminals has been established. DNA fingerprints are also used for immigration cases to prove genetic parenthood. There have been disputes about the probability of two individual's DNA fingerprints matching, and calculations have recently been revised, but there is still a low probability of such fingerprints matching (when the size of the fragments of DNA made after digestion with the restriction enzymes, are the same in two individuals).

The question of fairness in the use of genetic information with respect to insurance, employment, criminal law, adoptions, the educational system and other areas must be addressed. In those countries with private medical insurance, some people may be put into high risk or uninsurable groups because of genetic factors (for example, high blood cholesterol, or family history of diseases). Some insurance companies and employers perform screening in the USA. Some legislation has been passed in the USA, such as the Americans with Disabilities Act in 1990, or the proposed Human Genome Privacy Act. However, there is only one solution, that is a national health insurance system with equal access to all and no discrimination based on disease risk factors that people have no choice over.

GENE THERAPY

Many genetic diseases may be able to be treated by correcting the defective genes, by gene therapy. Gene therapy is a therapeutic technique in which a functioning gene is inserted into the somatic cells of a patient to correct an inborn genetic error or to provide a new function to the cell. There are a number of human gene therapy trials currently underway, for several different diseases including several cancers. There have been over a hundred human gene transfer or therapy trials on nearly a thousand patients approved in numerous countries. It is still an experimental therapy, but if it is safe and effective, it may prove to be a better approach to therapy than many current therapies, because gene therapy cures the cause of the disease rather than merely treating the symptoms of a disease. Also, many diseases are still incurable by other means.

Currently such gene therapy is not inheritable, we need to have much wider discussion about the ethics and social impact before we start inheritable gene therapy. However, non inheritable gene therapy to treat patients involves similar issues to any other therapy, and if it is safer and more effective, it should be available to patient's who consent to it. There are many approaches being developed and we can expect rapid introduction of these techniques, because of more than 20 years of preparatory experiments on animals, and the success of some of the current clinical trials.

We can distinguish gene therapy from genetic engineering for ethical debate, by the therapeutic motive. It is better to leave the term human genetic engineering for genetic intervention which is not aimed at direct therapy. However, we may still think of cases where indirect medical benefit can be given, such as improvement of the efficiency of the immune system by genetic means or genetic vaccination. Cosmetic uses of medicine are commonly available from the private medical sector, and it may be difficult to limit the cosmetic use of gene therapy in the future when cheap vectors and delivery systems have been developed. Public opinion supports the use of gene therapy for treating disease, in India as in other countries.

POPULATION GENETICS

Population genetics is a discipline studying genetic variation in defined populations, including relevant aspects of population structure and geographic variability of DNA sequences and their frequencies. Their changes in time and space are controlled by evolutionary factors, including mutation, natural selection (i.e. differential mortality and fertility of genetic types), drift (stochastic fluctuation tied to the demographic size of populations) and migration. Population genetics deals with the characteristics of genes within a population as opposed to the description of the genes in a particular individual.

The biological relationships of human population groups is of broad interest to the understanding of human history. Classical studies of genetic diversity have been dealing with antigen, protein and enzyme polymorphisms, for example HLA or blood groups. Modern genetic studies are based on the molecular analysis of DNA polymorphism. Classical studies look at expressed sequences, which represent less than 10% of the genome, whereas molecular genome diversity studies are mainly concentrated in parts of the genome that are often not expressed at the phenotypic level.

Isolated populations are the main source for observation of genetic forces acting in human evolution. While gene frequency across different populations varies for specific diseases; the total effect of genes on mortality appears similar. Comparative studies can be made on the differences in the mortality and fertility as well as on anthropometric data between consanguineous and non-consanguineous groups. They can be used to test genetic susceptibility using polymorphic markers in different communities, which can also examine genetic susceptibility to environmental agents. This makes possible the prevention of certain multifactorial diseases by careful avoidance of exposure to environmental agents.

Consanguinity and large family size are very interesting for population genetic studies, but studies of consanguineous marriages of families have the potential to raise numerous ethical and social issues. There are opportunities for genetic epidemiological research in countries where consanguineous marriage is a long-standing tradition. Such research projects can use data from many sources, for example, birth and death records, family register books, or anthropological or medical surveys, and socio-economic data in order to test inbreeding effects.

The first genetic studies in populations came from surveys all over the world looking at frequencies of single gene diseases. Others were conducted on sampled individuals from target populations. Based on these studies, mass screening programmes for particular disease genes were adopted e.g. in Cyprus for thalassemia, or in many countries for PKU in newborns. Genetic screening and genetic testing have been discussed by numerous persons and organisations over the last twenty years.

Many of the issues in current genetic screening programmes are relevant to population genetics research, but some significant issue are different, such as the notion of group consent. Moreover, while at present we are focusing on research we should also foresee the applications and benefits flowing from the human genome project that could well apply to whole populations. Therefore the ethical considerations must be carefully treated, since whole groups of asymptomatic individuals are the targets rather than single individuals who come forward themselves. More recently, molecular biology has enabled geneticists to work out the spatial and temporal variation of gene frequencies. Several projects have started independently in different countries others, such as the human genome diversity project (HGDP), proceed as an international addition to the human genome project (HGP).

HUMAN GENOME DIVERSITY PROJECT

One example of population genetics research is the Human Genome Diversity Project (HGDP) described by L. Cavalli-Sforza as"an international anthropology project that seeks to study the genetic richness of the entire human species". The name comes from a proposal in 1991 in the journal Genomics to make a systematic study of the genetic diversity of human populations. However, like the Human Genome Project, it shares a much older origin in the work of population geneticists over many decades. The Human Genome Organisation (HUGO) responded to the 1991 proposal in the journal Genomics by establishing an ad hoc committee to develop the global project. The HGDP is being developed under the auspices of HUGO to promote global involvement and coordination.

The scientific aims of the HGDP stated in the 1994 HUGO Summary Document are:

- "To investigate the variation occurring in the human genome by studying samples collected from populations that are representative of all of the world's peoples,"
- "And ultimately, to create a resource for the benefit of all humanity and for the scientific community world-wide.

The resource will exist as a collection of biological samples that represents the genetic variation in human populations world-wide and also as an open, long-term, genetic and statistical database on variation in the human species that will accumulate as the biological samples are studied by scientists from around the world."

The main scientific value of the HGDP is:

- Deepening our understanding of human history and identity.
- Gaining knowledge about the environmental and genetic factors involved in predisposition and resistance to disease, so-called genetic epidemiology.
- Encourage the development of local laboratories where the collection of genetic samples will be collected and analysed.

Even though to date there have been numerous studies on the development of culture, language and population genetics and some consistency between genetic, cultural and linguistic observations has been found, a survey of more populations in a more systematic way will extend what we already know and test current theories. Linguistic differences suggest there are about 5,000 population groups in the world. In the short term the HGDP will attempt to study about 500 of these populations. Even if some populations refuse to enter the project, there are still many other populations that could be surveyed. It is expected, then, that the project will be able to obtain samples from a large number of willing populations. If funding does not permit such wide sampling there is still scientific merit in collecting data from a smaller number of populations. The HGDP initially planned to centralise the collection of samples from isolated populations, some of which are already under investigation in population genetics research. This led to fears among some members of indigenous groups that the knowledge could be used for further ostracisation. However, the HGDP is now moving away from both the idea of central control to regional control and from the focus on indigenous populations, to include all populations. Personal anonymity would be maintained by not having the names of individuals in the central repository, and by observance of established privacy rules.

The establishment of cell lines allows maintenance of a permanent record of the DNA of individuals of a population. At least two independent and physically separate collections in different countries should be kept, to maintain the resource. The HUGO HGDP committee has said that access will be free, with some compensation for maintenance costs. Any data would be shared back into the main database, which would also include computer databases of genetic map and sequence data. There are also efforts to develop less expensive storage and microsatellite marker techniques that can be used in local laboratories that have limited resources, to ensure their fuller participation.

ETHICAL ISSUES OF POPULATION GENETICS RESEARCH

Given the transnational nature of population genetics studies, the issues raised were the subject of a study by the UNESCO International Bioethics Committee. The report has been available on the Internet since completion in November 1995.

Key points of the UNESCO report were:

- No endorsement of a particular population genetic project;
- Call for establishment of a separate ethical committee that is available to all population genetics researchers;
- Discussion of variety of ethical, ethnic and social issues.

The ethics of population genetics should be formulated with reference to the minimal agreed human values as expressed in international human rights law. Central to our human rights obligations is the promotion of"respect for, and observance of, human rights and fundamental freedom for all without distinction as to race, sex, language, or religion.".

The importance of obtaining consent from a study population is well recognised (even if not always practiced). Population genetics studies should be conducted by personnel with the appropriate qualifications, but only after informed consent has been given for any general medical examinations, and removal of blood samples or other bodily samples. The doctrine of informed consent is applied to both medical treatment and research. Before a person is asked to consent to any sampling or treatment they must be provided with certain information. This information includes (in language the patient can understand) a description of the procedure - which is generally easy, and should be risk free if accepted medical procedures are used for sampling, and a description of the risks and benefits of the resultant information.

Even when correct information is carefully presented in culturally appropriate ways, it cannot be guaranteed that it has been understood. The ethical obligations that are achievable include accurate delivery of information together with the disclosure of relevant risks and benefits to the individuals and communities involved, in language that is accessible to the potential research subjects, considering cultural and religious needs and aspirations of particular communities. There is an obligation or duty on the part of scientists to properly inform potential participants.

It is important to identify the most appropriate persons with whom to communicate, the persons from whom clearance should be obtained, and the appropriate content and media of communication. If the research methods involve the use of saliva, skin, hair, or blood samples it is necessary to ensure that the collection of these body samples does not violate cultural norms and concepts relating to the human body and its functions. Of course the most difficult group consent question is who can give group consent for the genome project itself - a question that no one has been able to answer.

As discussed in following papers, various groups of indigenous peoples have expressed their irritation with past population genetics research which they claim has been conducted without prior consultation and in a way where consent was obtained in terms inconsistent with their cultural norms. Practitioners of contemporary science do not always understand that the goals and aspirations of scientific projects may not always coincide with the goals and aspirations of particular cultures. It may not be ethically acceptable to some people to cooperate in the collection of saliva, skin, hair and blood samples for the purpose of storage and the establishment of"transformed cell lines", samples which would be basically identical to the individual of origin which are then made available for study to scientists around the world. Therefore many representatives of indigenous peoples have expressed strong concerns about the HGDP.

The planners of the HGDP got off to a bad start with misunderstandings and fears widely expressed among indigenous peoples. Because the HGDP was planning to collect blood samples some groups called the HGDP the"Vampire project", while other groups were angry because they believed that they were possible target populations even though no community representatives had been contacted about the Project. The Mataatua Declaration on Cultural and Intellectual Property Rights of Indigenous Peoples of June 1993 is a call for a halt to the HGDP until its impact has been discussed. Article 3.5. of the Declaration calls"for an immediate halt to the on-going 'Human Genome Diversity Project' (HUGO) until its moral, ethical, socio-economic, physical and political implications have been thoroughly discussed, understood and approved by indigenous peoples".

The Declaration is actually not anti-science, and includes a call for involvement in scientific research, recommendation 2.11,"Ensure current scientific environmental research is strengthened by increasing the knowledge of indigenous communities and of customary environmental knowledge".

In fact the HGDP included all populations, not only indigenous populations. Since that time, the HGDP goals have shifted somewhat, but there has been a series of Declarations directed against this project. The HUGO HGD Summary Document includes ethical guidelines which do address the question of participation, consent, and commercialisation (HGD, 1994); as does the HUGO Code of Ethics. The Human Genome Organization (HUGO) Ethics Committee Code of Ethics for genetic sampling stresses similar issues to those raised by the UNESCO subcommittee.

It starts from the same principles: Recognition that the human genome is part of the common heritage of humanity; Adherence to international norms of human rights; Respect for the values, traditions, culture, and integrity of participants; and acceptance and upholding of human dignity and freedom. The HUGO - ELSI Committee Code of Ethics was endorsed by HUGO in March 1996. It recommends principles for genetics research based on: scientific competence, understandable communication (two way), consultation with participants, free and informed consent, respect the choices of those involved, confidentiality, international collaboration, avoid conflict of interest, undue inducement through compensation, and continual review of the research.

Nevertheless, the controversy continues. In February, 1995, a forum of Indigenous Peoples in Asia issued a statement to the European Parliament in which they strongly opposed the HGDP and called for it to be stopped. The Beijing Declaration of Indigenous Women formulated at the United Nations Fourth World Conference on Women demanded"that the Human Genetic Diversity Project be condemned and stopped"and that their"intellectual community rights"be recognised.

There are numerous other ethical issues which are common to the HGP and the HGDP. Many issues relate to individual privacy and possible abuse of genetic data by insurance companies and employers. DNA collected from population groups would present analogous problems relating to the adequate protection of privacy; but the meaning of privacy could vary across cultures. Furthermore, it is important to note that in population genetics research confidentiality issues have to be considered at the community level as well as at the individual level.

One way of ensuring confidentiality for individuals would be not to collect or keep identifying information in the central repository. This would, however, limit the scope of research that could be done. Confidentiality for individuals in human population genetic research may be protected to a certain extent through coding and anonymity, with strong safeguards to protect the identifying information that is kept in the central repository. If all members of a community, or a population group, were found to have a gene that predisposes them to a common late onset disease, for example, could that information be protected? Health insurance companies could cancel or refuse health insurance to a population, in the same way that health insurance has been denied to individuals or members of a family in countries that do not prohibit genetic discrimination by law. In several countries, these cases have led to legislation on genetic privacy. Although the principle of confidentiality is included in the Outline of the UNESCO Declaration on the Protection of the Human Genome, further national efforts to protect against such abuses may be required.

Patenting was discussed in a separate section above. While there could also be provision for a one-time gift of cells or blood with no conditions, as is found in some tissue donation forms for blood and body tissues, can one individual sign away commercial rewards to future research knowledge for the population to which they belong? Financial returns are not the only form of benefits of research results which could be returned to subjects of research. At the individual level, the results of physical examinations and clinical diagnosis and options for treatment are sometimes communicated to each participant in the local languages through local health authorities and doctors. The provision of health and medical care should be appropriate to the cultural and social context of the community and should be sustained. In this, the principles of primary health care as contained in the Alma-

Ata Declaration of 1978 should serve as a good set of guidelines to follow. At the community level, the health data could be utilised for the improvement of local community health. Thus, benefits should also flow back to the groups and communities in the form of contributing to the formulation and implementation of local and national health care policies that would enable communities to better their positions. These policies, as well as the health care services which are offered, should of course be decided upon by the communities.

Some challenging implications may arise from the better understanding of human history that population genetics research could provide. The new knowledge from such research could be used to educate people of indigenous groups that would help protect their interests. On the other hand, new knowledge of human group evolution and the relationships of particular groups to others, may challenge existing world-views. Some population groups have strong beliefs in mythologies or cosmogonies which explain group origins and identity, and the return of data that challenges the accepted beliefs could be a delicate issue. One fear that has been expressed about population genetics research is that access to and knowledge of a community's complete genetic make-up makes it theoretically possible to devise cheap and targeted biological weapons trained solely on that community. Given that greater genetic diversity exists within any particular population group than among population groups, it is highly unlikely that specific genetically-based genocidal weapons could be developed for specific population groups.

Public attitudes towards population genetics are often based on social ideologies, racism, and eugenics and can well lead to stigmatisation and genetic reductionism. Population studies in the past have shown that most of the genetic diversity is to be found within every race or population, and if this is further confirmed to be true the typological classification of humans into different"races"is scientifically invalid. Nevertheless"racism"as an ideology and as an attitude is a human reality. Eugenics is a word coined by Francis Galton in 1883 meaning good genes. He refered to the"science"of improving human stock by giving"the more suitable races or strains of blood a better chance of prevailing speedily over the less suitable". A danger is that genetic findings may be understood as the only story of human, biological, and cultural evolution. People believe genetic explanations of human behaviour, as can be seen in the scientific and popular press over the past decade. Stigmatisation may occur when population groups in which there is a high incidence of genetic disease are selected for scientific investigation thereby drawing attention to their genetic differences.

Care needs to be taken that targeted groups do not become stigmatised in some way simply because they are of scientific curiosity, or because they are more frequently studied and more is known about them they seem to have a greater predisposition to disease. Such stigmatisation can lead to unjust discrimination.

"Genetic reductionism"means judging individuals with reference only to their genetic inheritance. Western science has frequently sought to explain the whole by a greater knowledge, by a part that is deemed to be most fundamental. Genetic reductionism is a misunderstanding of science and also represents a threat to those mythologies or cosmogenies which are different from the dominant world cultures.

One of the fundamental points of opposition of indigenous groups towards genetic studies of human history is that the results may contradict indigenous people's views of oral and traditional history, and the meaning of genes and genealogy. For example, Maori people have two words to describe the human gene, one meaning"life spirit of mortals (Iratangata)and the other genealogy (Whakapapa), which connects Maori with themselves and others.

The gene and genome are not the property of individuals but rather are part of the heritage of families, communities, tribes and entire indigenous nations. In this regard the UNESCO position on the human genome being part of the common human heritage is more compatible with the views of indigenous persons, than the view discussed in the patent section of those seeking patents on genes.

Bibliography

Claudio, A.: *Proteomics and Nanocrystallography*, Cambridge Cambridge University Press, 1999.

Claudio, N.: *Nanobiotechnology and Nanobiosciences*, Berkeley University of California Press, 2005.

David, P.: *Biotechnology: Applying the Genetic Revolution*, Princeton Princeton University Press, 2001.

David, W.: *Nanobiotechnology Protocols*, Providence: CRC-Press 2001.

Dekker, M.: *Nano-surface Chemistry*, USA: Oxford University Press 2002.

Donald, K.: *Nanobiotechnology of Biomimetic Membranes*, USA Oxford University Press, 1998.

Donald, M.: *Leading Edge Nanotechnology Research Developments* New York: Chapman and Hall, 2000.

Dubowski, J. J.: *Photon-based Nanoscience and Nanobiotechnology* New York: Springer, 1998.

Edward, S.: *Nano-and Micro-electromechanical Systems: Fundamental of Nano*, Princeton: Princeton University Press, 2001.

Ehud, G.: *Plenty of Room for Biology at the Bottom: An Introduction to Bionanotechnology*, New York: Springer, 1995.

Johann, S.: *Nano-bioethics: Ethical Dimensions of Nanobiotechnology* Princeton: Princeton University Press, 2004.

Kozyrev, S.: *Silicon Versus Carbon: Fundamental Nanoprocesses Nanobiotechnology*, Berkeley: University of California Press 2004.

Mahlich, C.: *Innovation and Technology in Korea: Challenges of a newly Advanced Economy*, Princeton: Princeton University Press, 2000.

Malanowski, N.: *Growth Market Nanotechnology: An Analysis of Technology and Innovation*, New York: Springer, 2002.

Marvin, V.: *Advances in Computers: Nanotechnology*, Princeton: Princeton University Press, 2000.

Niemeyer, M.: *Niemeyer Nanobiotechnology more Concepts and Applications*, New York: Springer, 2005.

Paolo, S.: *Scanning Probe Microscopies beyond Imaging: Manipulation of Molecules*, Berkeley: University of California Press, 2000.

Philippe, H.: *Nanoscience: Nanobiotechnology and Nanobiology*, New York: Chapman and Hall, 2002.

Ralph, S.: *Nanoscale Technology in Biological Systems*, Providence: CRC-Press, 2004.

Sandro, C.: *Nano-bio-sensing*, USA: Oxford University Press, 1999.

Swapan, K.: *Glioblastoma: Molecular Mechanisms of Pathogenesis and Current Therapeutic*, Cambridge: Cambridge University Press, 2002.

Thomas, M.: *Fundamentals of Tissue Engineering and Regenerative Medicine* USA: Oxford University Press, 2000.

Werner, M.: *The nano-micro Interface: Bridging the Micro and Nano world*, Providence.: CRC-Press, 2000.

Yount, L.: *Biotechnology and Genetic Engineering*, Berkeley: University of California Press, 2002.

Zoltan, D.: *Colloids for Nano-and Biotechnology*, Cambridge: Cambridge University Press, 2000.

Index